A Concise Guide to Statistical Analyses

A Concise Guide to Statistical Analyses

Using Excel®, SPSS®, and the TI-84 Calculator

SHERRI L. JACKSON
Jacksonville University

WADSWORTH
CENGAGE Learning·

Australia • Brazil • Japan • Korea • Mexico • Singapore • Spain • United Kingdom • United States

WADSWORTH
CENGAGE Learning·

A Concise Guide to Statistical Analyses:
Using Excel®, SPSS®, and the TI-84
Calculator
Sherri L. Jackson

Publisher-in-Chief: Linda Ganster

Publisher: Jon-David Hague

Acquisitions Editor: Timothy Matray

Assistant Editor: Lauren K. Moody

Editorial Assistant: Nicole Richards

Marketing Program Manager: Janay Pryor

Art and Cover Direction, Production
Management, and Composition:
PreMediaGlobal

Manufacturing Planner: Karen Hunt

Rights Acquisitions Specialist:
Roberta Broyer

Photo Researcher: Jeremy Glover

Cover Image: ©kentoh/Shutterstock

For product information and technology assistance, contact us at
Cengage Learning Customer & Sales Support, 1-800-354-9706.
For permission to use material from this text or product,
submit all requests online at **www.cengage.com/permissions**.
Further permissions questions can be e-mailed to
permissionrequest@cengage.com.

Library of Congress Control Number: 2012935029

ISBN-13: 978-1-133-31553-7

ISBN-10: 1-133-31553-4

Wadsworth
20 Davis Drive
Belmont, CA 94002-3098
USA

Cengage Learning is a leading provider of customized learning solutions with office locations around the globe, including Singapore, the United Kingdom, Australia, Mexico, Brazil, and Japan. Locate your local office at **www.cengage.com/global**.

Cengage Learning products are represented in Canada by
Nelson Education, Ltd.

To learn more about Wadsworth, visit **www.cengage.com/wadsworth**.

Purchase any of our products at your local college store or at our preferred online store **www.cengagebrain.com**.

All Excel screen shots in this edition are used with permission from Microsoft Corporation. All rights reserved.

All SPSS screen shots in this edition are courtesy of IBM.

Printed in the United States of America
1 2 3 4 5 6 7 16 15 14 13 12

Table of Contents

Preface

The intent of this manual is threefold. First, the manual covers a limited number of statistical analyses, those primarily taught in an undergraduate statistics class. Second, it provides details on how to use not only SPSS to conduct these analyses but also explanations of how the same analyses can be conducted using Excel or the TI-84 calculator (resources to which undergraduates are far more likely to have access and with which they are more likely to be familiar). Third, the manual simplifies the presentation of how to accomplish each statistical analysis using these tools, and thus brings the presentation to a level that most undergraduates will be able to follow easily.

Consequently, if an institution does not have the resources to purchase SPSS, students can still begin to analyze data using Excel (to which students at almost all schools have access) or the TI-84 calculator (which students can easily carry to class if laptops or desktops are not part of the classroom setting). The manual, therefore, will have a much wider application than many of the traditional manuals that cover either SPSS or Excel, but not both. In addition, because students are more familiar with Excel and because the Excel output is not nearly as detailed as that from SPSS, students often find it easier to conduct analyses and interpret output using Excel. Likewise, they often find it easier to begin by using Excel and then transfer that knowledge to SPSS. The inclusiveness of this manual thus makes it more beginner friendly. In addition, because the use of Excel, SPSS, and the TI-84 calculator are all covered, the manual is a more complete resource for students and faculty.

Many faculty members are unaware that both Excel and the TI-84 calculator are fairly comprehensive in terms of the statistical analyses they are capable of producing. Excel calculates most of the basic inferential statistics that SPSS does, including ANOVA (one-way randomized and repeated measures and two-way), correlation, and regression analyses. Additionally, the TI-84 cannot only calculate descriptive statistics but also several inferential tests (e.g., t tests and one-way

ANOVA), correlation, and regression analyses—a quite impressive range for something that is so easily carried to class.

The format of this manual comprises chapters divided into modules that present information in digestible chunks. Each module first introduces a statistical problem and then addresses how the problem can be solved using each of the three tools (assuming the tool is capable of conducting a particular analysis). Therefore, for example, two-group designs are introduced with one problem that would require an independent-groups *t* test and another that would require a correlated-groups *t* test. For each type of test, the problem is explained, and the data are provided. A description then follows of how to conduct this test first using Excel, then SPSS, and finally the TI-84 calculator. When a description is missing—for example Excel does not calculate chi-square tests using the Data Analysis Toolpak—it means that that resource does not calculate that test. Visuals (screen captures) are provided for both Excel and SPSS problems so that students can see exactly what the computer screen should look like when they conduct each test. The final output screen is also provided so that students can check their output against that provided in the text. In addition, the relevant data lines from the output screen are discussed so that students know exactly to what they should be paying attention from the output. Lastly, each module has module exercises so that students can practice using the resources to calculate various statistics. The answers to the module exercises appear in the Appendix.

In summary, the key features of this manual are:

- Simplicity of presentation
- Coverage of three resources rather than just one
- Statistics that more closely match typical undergraduate courses
- Chapters that are divided into modules so that material is presented in smaller chunks
- Module exercises with answers in the Appendix that allow students to apply what they have learned

Sherri L. Jackson

CHAPTER 1

✳

Getting Started

In this first chapter, we'll learn how to enter data into Excel, SPSS, and the TI-84 calculator. This process includes using the data editors and naming variables. In addition, Module 2 will illustrate how to create graphs (bar graphs, histograms, and frequency polygons) and tables using Excel and SPSS.

Module 1

✳

Starting and Using Excel, SPSS, and the TI-84 Calculator

USING EXCEL 2007 OR 2010

Installing the Analysis ToolPak

Before you begin to use Excel to analyze data, you may need to install the Data Analysis ToolPak. For Excel 2007, this can be accomplished by launching Excel and then clicking on the Microsoft Office icon at the top left of the page. At the bottom of the drop-down menu, there is a tab labeled **Excel Options**.

Click on **Excel Options**, and a dialog box of options will appear. On the left-hand side of the dialog box is a list of options; click on **Add-Ins** to open a pop-up window. The very top option in the pop-up window should be **Analysis ToolPak**. Click on this and then **GO**.

A dialog box in which **Analysis ToolPak** is the first option will appear. Check this and then click **OK**.

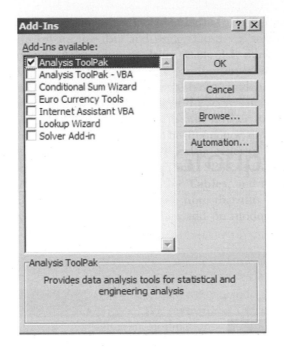

The ToolPak should now be installed on your computer.

For Excel 2010, the process is similar. Begin by clicking on the **File** tab at the top left of the screen (this tab replaces the Microsoft Office icon from 2007).

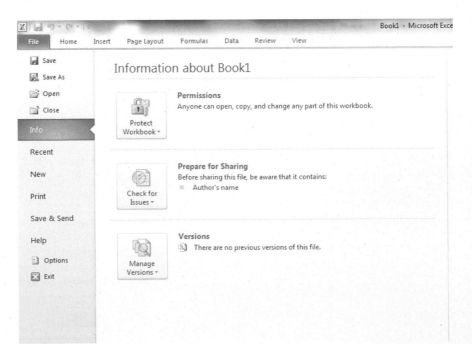

Next, click on **Options** toward the bottom of the drop-down menu. From here the process is similar to that for 2007. A dialog box of options will appear. On the left-hand side of the dialog box is a list of options; click on **Add-Ins** to open a pop-up window. The very top option in the pop-up window should be **Analysis ToolPak**. Click on this and then **GO**. A dialog box in which **Analysis ToolPak** is the first option will appear. Check this and then click **OK**.

Launching Excel

When you open Excel, you will be presented with a spreadsheet.

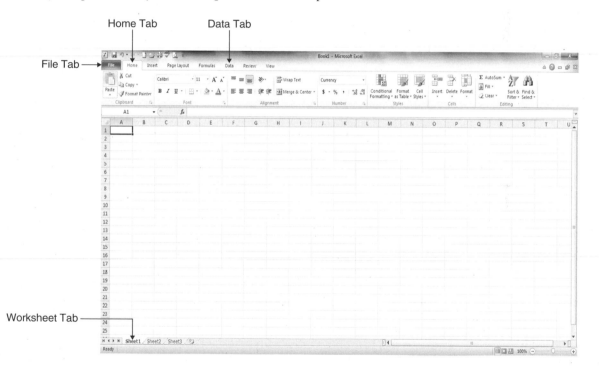

You can see in the preceding worksheet where the **File** tab is (in Excel 2007, the Office Button is here), along with the **Home**, **Data**, and **Worksheet** tabs. The spreadsheet has a series of columns and rows. The columns are labeled with letters, and the rows with numbers. A cell is the point at which a column and row intersect. Thus, Cell A1 is highlighted in the preceding screen capture. We typically refer to cells with the column letter first and the row number second (e.g., D5). We will be entering data into the cells for the analyses we'll be conducting using Excel in subsequent modules. In addition, we'll be using the Data Analysis ToolPak for most of our analyses. The ToolPak can be accessed when the **Data** ribbon is active (see the following screen capture).

When the **Data** ribbon is active, as in the preceding image, you can see the **Data Analysis** option at the upper right of the screen. This option indicates that the ToolPak has been installed properly. If the **Data Analysis** option is not visible at the top right of the ribbon, then the ToolPak was not installed properly (see previous instructions on installing the ToolPak).

Entering Data

We can enter data into Excel by simply entering the numbers in the cells. For example, in the following image, I've entered some annual salary data for a small business. You can see that I labeled the data **Salary** in the first cell and then entered the numerical data in a column below the heading.

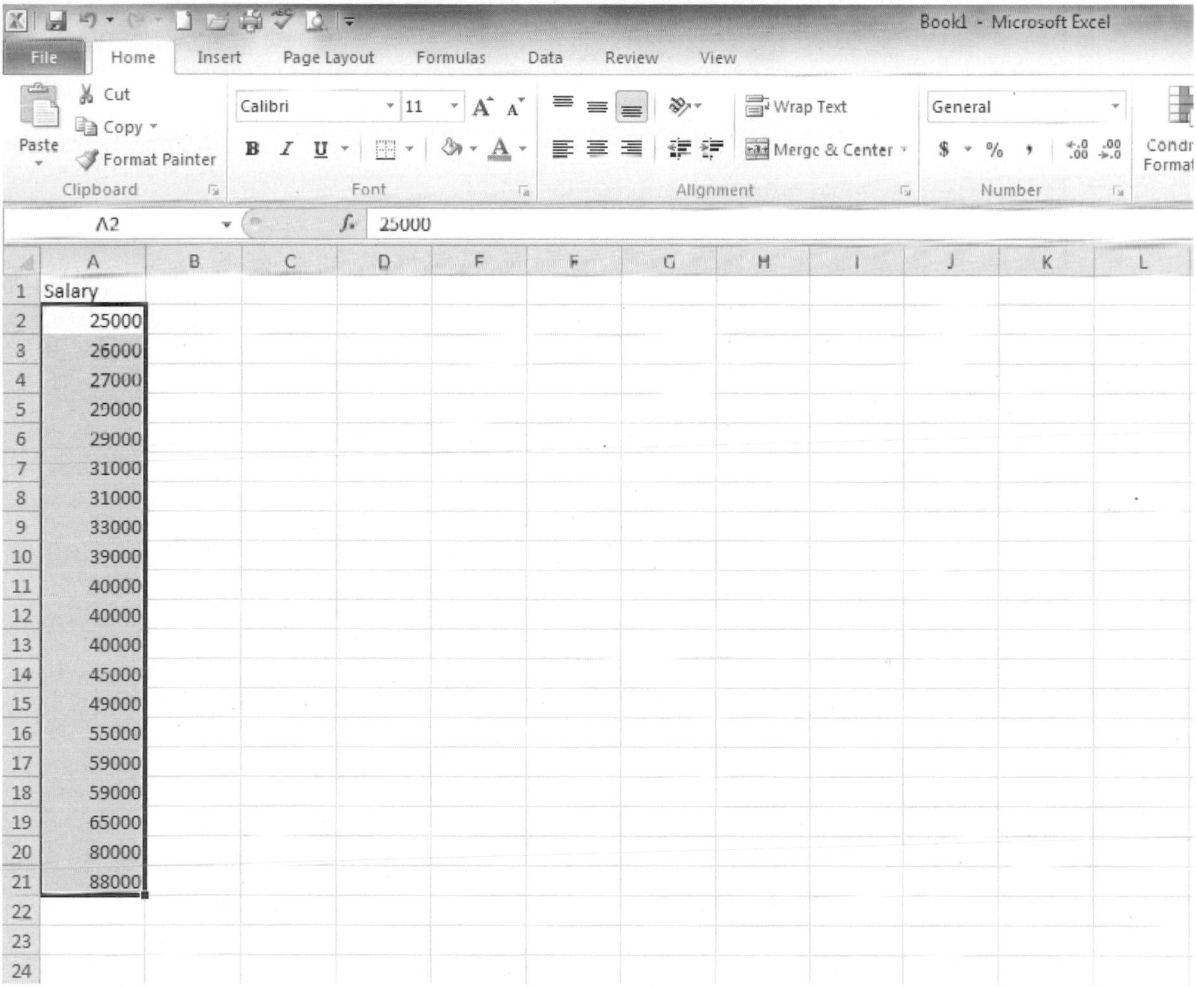

If I want the data to have a different appearance—for example, to indicate that it is currency—I can format the number type through the **Home** ribbon, **Number** window. To begin, highlight the data in Column A (see the preceding image). Then activate the **Home** ribbon by clicking on it (this has been done in the preceding image). You should see the **Number** window in the top right-hand side of the **Home** ribbon in the preceding image.

You can activate the **Number** window by clicking on the arrow in the bottom right of the window; you will receive a dialog box as in the following screen capture that will allow you to format the way you want the numerical data to appear in the cells. Because these data are salaries, let's choose **Currency** and then click **OK**.

The data in Column A now appear as currency as can be seen in the following image.

USING SPSS 18 OR 19

To begin using SPSS, launch the SPSS program. When you initially start the program, you may receive a startup window such as the one below. Unless otherwise instructed by your professor, click **Cancel**. (Please note: SPSS 18 was marketed as PASW Statistics; it returned to being named SPSS in version 19.)

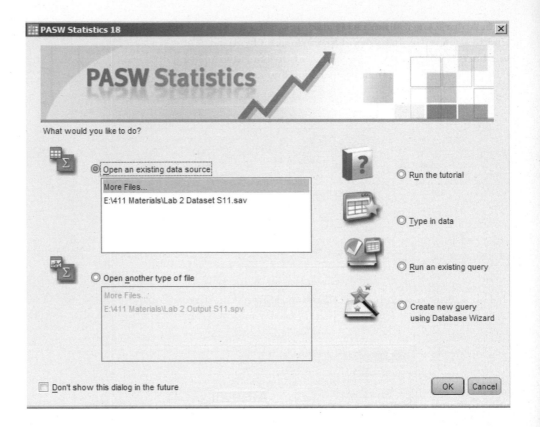

You should now have a view of the data spreadsheet, or the Data Editor, similar to the following image.

The main menu in SPSS runs across the top of the Data Editor and gives you the options from **File** on the far left to **Help** on the far right. We'll mainly be using the **Analyze** option to statistically analyze data. You can see the drop-down menu for the **Analyze** tab in the following image.

The icons that appear below the main menu provide some shortcut keys to allow you to do such things as Open Files, Save, Insert Cases, Insert Variable, and so on.

Entering Data

To enter the data into an SPSS spreadsheet, launch SPSS and enter the data into a column, as in the following spreadsheet in which I have entered the salary data from the previous Excel example. Given that I had already entered the data into Excel, I simply cut and pasted the data column into SPSS, minus the **Salary** heading.

Notice that the variable is simply named VAR0001. To rename the variable to something appropriate for your data set, click on the **Variable View** tab at the bottom left of the screen. You will see the **Variable View** window active in the following image.

Type the name you wish to give the variable in the first column (labeled **Name**). The variable name cannot have any spaces in it. Because these data represent salary data, we'll type in **Salary**. You can also see that we can format this variable while the **Variable View** window is active by specifying the type of data, the width of the data, the number of decimals, and so on. To get back to the original spreadsheet, highlight the **Data View** tab at the bottom left of the screen. In the **Data View** window, we can now see that the **Salary** heading appears at the top of the column.

File	Edit	View	Data	Transform	Analyze	Graphs	Utilities	Add-ons	Window	Help

1 : Salary		25000.00										
	Salary	var	var	var	var	var	var	var	var	var	var
1	25000.00										
2	30000.00										
3	30000.00										
4	32000.00										
5	33000.00										
6	33000.00										
7	35000.00										
8	35000.00										
9	35000.00										
10	35000.00										
11	37000.00										
12	37000.00										
13	40000.00										
14	40000.00										
15	40000.00										
16	45000.00										
17	45000.00										
18	52000.00										
19	55000.00										
20	56000.00										
21	65000.00										
22	90000.00										

If I had not had data to paste into the Data Editor, I could have entered the data by highlighting each cell and typing in the data.

USING THE TI-84 CALCULATOR

Entering data into the TI–84 calculator is considerably easier than entering it into Excel or SPSS. With the calculator on, press the **STAT** key. **EDIT** should be highlighted at the top left of the screen along with **1: Edit** in the first row under the menu. Press the **ENTER** key at the bottom right of the calculator. You should now have a spreadsheet with six columns labeled **L1** through **L6** (you may only be able to see L1 through L3, but if you scroll to the right, columns L4 through L6 become available). To enter data into L1, move the cursor to the first position under L1 and type in the data. Press **ENTER** after you type each number. Your data should now be entered into L1 and be displayed in the column under L1.

To clear the data from a column, highlight the column heading (for example, L1) and press the **CLEAR** key on the calculator. Never press the **DEL** key after highlighting a column heading because this action will delete the entire column. However, if you make this mistake, you can recover the column by pressing the **STAT** key, scrolling down to **5: SetUpEditor**, and pressing **ENTER**. You will receive the command **SetUpEditor** with a flashing cursor. Press **ENTER** on the calculator, and you should receive the message **DONE**. Your variable should now be restored.

Module 2

✳

Graphs and Tables

W̲e will discuss two methods of organizing data: graphs and tables.

GRAPHS

Several types of pictorial representations can be used to represent data. The choice depends on the type of data collected and what the researcher hopes to emphasize or illustrate. The most common graphs used by psychologists are bar graphs, histograms, and frequency polygons (line graphs). Graphs typically have two coordinate axes: the x-axis (the horizontal axis) and the y-axis (the vertical axis). Most commonly, the y-axis is shorter than the x-axis, typically 60% to 75% of the length of the x-axis.

Bar Graphs and Histograms

Bar graphs and histograms are frequently confused. When the data collected are on a nominal scale or when the variable is a *qualitative variable* (a categorical variable for which each value represents a discrete category), then a bar graph is most appropriate. A *bar graph* is a graphical representation of a frequency distribution in which vertical bars are centered above each category along the x-axis and are separated from one another by a space, which indicates that the levels of the variable represent distinct, unrelated categories. When the variable is a *quantitative variable* (the scores represent a change in quantity) or when the data collected are ordinal, interval, or ratio in scale, then a histogram can be used. A *histogram* is also a graphical representation of a frequency distribution in which vertical bars are centered above scores on the x-axis; however, in a histogram the bars touch each other to indicate that the scores on the variable represent related, continuous values. In both a bar graph and a histogram, the height of each bar indicates the frequency for that level of the variable on the x-axis. The spaces between the bars on the bar graph indicate not only the qualitative differences among the categories but also that the order of the values of the variable on the x-axis is

15

TABLE 2.1   Political Affiliation for a Distribution of 30 Individuals
© Cengage Learning 2013

Affiliation	Frequency
Democrat	12
Green	1
Independent	4
Republican	11
Socialist	2

arbitrary. In other words, the categories on the x-axis in a bar graph can be placed in any order. The fact that the bars are contiguous in a histogram indicates not only the increasing quantity of the variable but also that the variable has a definite order that cannot be changed.

Let's use the data from Table 2.1 for a hypothetical distribution of the frequencies of individuals who affiliate with various political parties to illustrate how to create bar graphs in both Excel and SPSS.

Using Excel to Create a Bar Graph

Begin by entering the data into an Excel spreadsheet, as follows. Please note that the column headings of Affiliation and Frequency are entered into the spread-sheet. Once the data are entered, highlight all of the data including the column headers.

Now select the **Insert** ribbon and then **Column** (please note that there is a **Bar** option for figures but that this produces horizontal bars, whereas the bars in a bar graph should be vertical). Select the top left option from the **Column** options (2-D clustered column chart). This should produce the following bar graph.

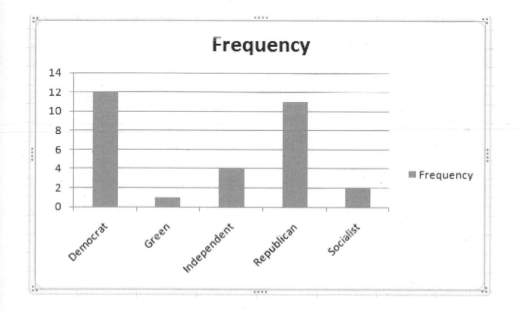

Notice that the different political parties are listed on the *x*-axis whereas frequency is recorded on the *y*-axis. Excel provides **Chart Tools** so that we can modify the appearance of a graph. For example, if you want the bar graph to conform to APA style, you could use **Chart Tools** to modify the appearance of the chart. To use **Chart Tools**, make sure that you have clicked on the chart in Excel after which the three ribbons under **Chart Tools**, (**Design**, **Layout**, and **Format**) will become accessible. Using these menus, you can, for instance, change the appearance of the chart to add **Axis Titles** (under the **Layout** ribbon), remove the horizontal **Gridlines** (under the **Layout** ribbon), or change the color of the bars (Excel uses blue as the default) by using the **Format** ribbon, clicking one of the bars, and selecting **Shape Fill**. After making these modifications, your chart will appear as follows.

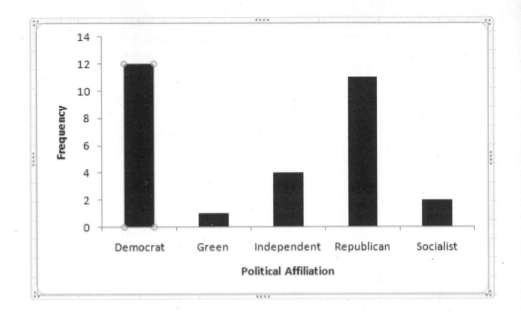

Please also note that although the political parties are presented in a certain order, this order could be rearranged because the variable is qualitative.

Using SPSS to Create a Bar Graph

We'll use the same data as in the previous example to illustrate how to use SPSS to create a bar graph. To begin, we enter the data into the SPSS spreadsheet. As with Excel, we use two columns, one labeled Affiliation and the other Frequency, as can be seen in the following screen capture.

File	Edit	View	Data	Transform	Analyze	Graphs	Utilities	Add-ons	Window	Help

4 : Frequency		11.00					
	Affiliation	Frequency	var	var	var	var	var
1	Democrat	12.00					
2	Green	1.00					
3	Independent	4.00					
4	Republican	11.00					
5	Socialist	2.00					
6							
7							
8							

Next, we click on the **Graphs** *menu* and then **Chart Builder**. From the Gallery menu on the bottom of the dialog box select Bar and then double click the first bar graph icon in the top row to produce the following dialog box.

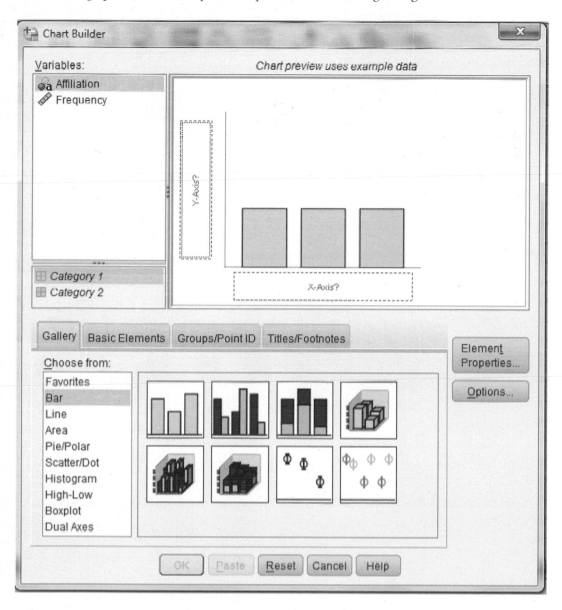

You can see that the two variables are listed in the top left **Variables** box. Drag the Affiliation variable to the *x*-axis box in the figure at the top right and then drag the Frequency variable to the *y*-axis box in the figure at the top right. The dialog box should now look as follows.

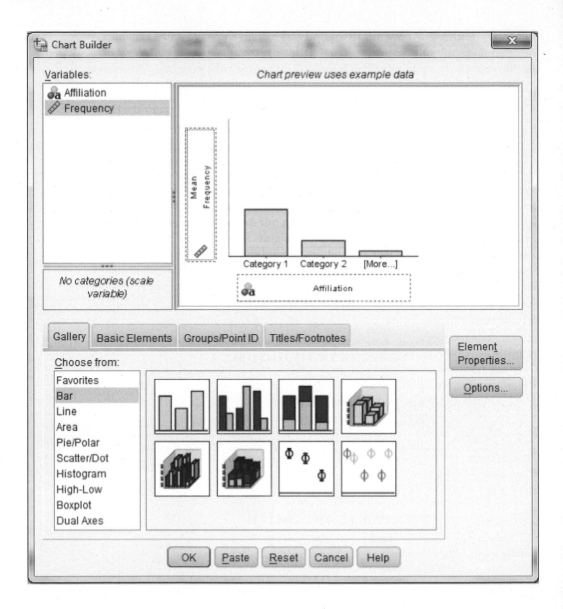

Next, click on the **Element Properties** box at the right-hand side of the dialog box to receive the following dialog box.

Highlight **Y–Axis 1 (Bar 1)** and change the name of the variable from Mean Frequency to simply Frequency, and then click **Apply** and **OK**. Finally, click **OK** in the original dialog box. SPSS will then produce an output file with the following bar graph.

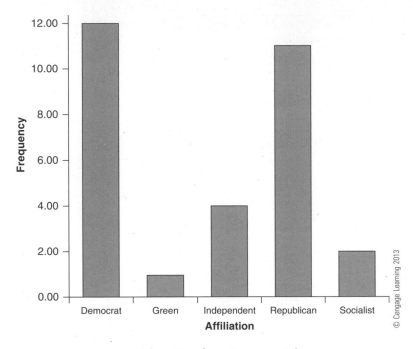

© Cengage Learning 2013

Using Excel to Create a Histogram

To illustrate the difference between a bar graph and a histogram, let's use the data from Table 2.2, which lists the frequencies of intelligence test scores from a hypothetical distribution of 30 individuals. A histogram is appropriate for these

TABLE 2.2   IQ Score Data for 30 Individuals
© Cengage Learning 2013

Score	Frequency
83	1
86	1
89	1
92	2
98	3
101	4
107	5
110	3
113	2
116	3
119	2
125	1
131	1
134	1

data because the IQ score variable is quantitative. The variable has a specific order that cannot be rearranged.

For Excel, begin by entering the data into an Excel spreadsheet, as follows. Please note that the column headings of Score and Frequency are entered into the spreadsheet. Once the data are entered, highlight only the Frequency data as is illustrated in the next screen capture.

Because Excel does not have a histogram option in which the bars in the graph touch, we'll have to use special formatting to create the histogram. Click on the **Insert** ribbon and then **Column**. Select the option at the top left corner, as we did when creating bar graphs. This should produce the following graph.

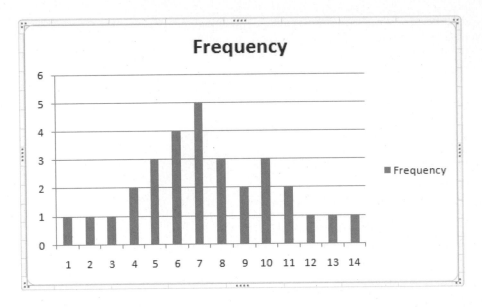

We'll begin editing the graph by removing the spaces between the bars. To do so, right click on any of the bars and select **Format Data Series** to produce the following pop-up window.

Move the **Gap Width** tab to zero as is indicated in the window and then close the window. Your figure should now more closely resemble a histogram. Now you can use the **Chart Tools** to modify your figure so that it more closely resembles what you desire. This should include axis labels on the x- and y-axes and changing the values on the x-axis to reflect the range of intelligence scores that were measured. To accomplish the latter, right click on a value on the x-axis and choose **Select Data** to produce the following pop-up window:

Select Data Source

Chart data range: ='Sheet1'!$A$1:$B$15

Switch Row/Column

Legend Entries (Series)

Add   Edit   Remove

Frequency

Horizontal (Category) Axis Labels

Edit

83
86
89
92
98

Hidden and Empty Cells   OK   Cancel

Click on the **Edit** window under **Horizontal (Category) Axis Labels**. You'll receive the following pop-up window.

Axis Labels

Axis label range:
=Sheet1!$A$2:$A$15   = 83, 86, 89, 92...

OK   Cancel

Highlight the IQ scores from the spreadsheet and they will be inserted into the **Axis label range:** box. Then click **OK**. Click **OK** a second time to close the original pop-up window. You can now use the **Chart Tools** to format your histogram so that it more closely resembles a graph appropriate for APA style. This would include adding axis labels to the x- and y-axes, changing the bars from blue to black, and removing the gridlines from the graph. After making these changes, your figure should look as follows:

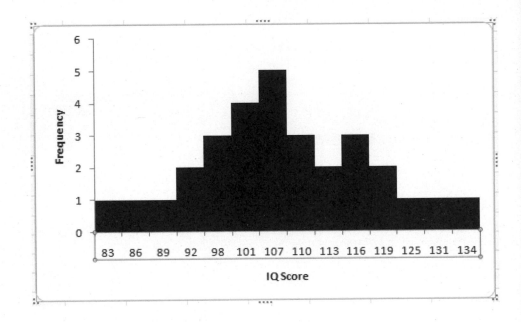

Using SPSS to Create a Histogram

Let's use the same data set as in the Excel example to create a histogram with SPSS. We'll enter the IQ score data into the SPSS spreadsheet, but in this case, each individual score is entered. This process is illustrated in the screen capture below in which all 30 scores have been entered into SPSS. (Please note that due to screen size constraints, the final four scores do not show in the screen capture. Thus, make sure you use the IQ data from Table 2.2.)

	IQscore	var	var
4	92.00		
5	92.00		
6	98.00		
7	98.00		
8	98.00		
9	101.00		
10	101.00		
11	101.00		
12	101.00		
13	107.00		
14	107.00		
15	107.00		
16	107.00		
17	107.00		
18	110.00		
19	110.00		
20	110.00		
21	113.00		
22	113.00		
23	116.00		
24	116.00		
25	116.00		
26	119.00		

Data View   Variable View

The variable was named IQscore using the **Variable View** screen, and it was designated a **Numeric** variable with the **Scale** level of measurement. From the **Data View** spreadsheet screen, select **Graphs** and then **Chart Builder** to receive the following dialog boxes.

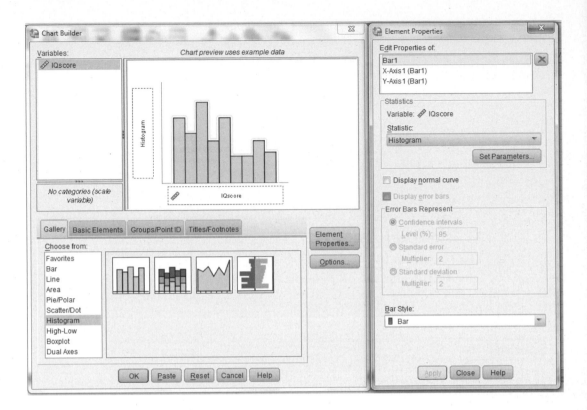

Select **Histogram** and then double click on the first example of a histogram. In the dialog box at the top left of the screen, click on **IQscore** and drag it to the *x*-axis box in the histogram on the right. Then, in the **Element Properties** box on the right, highlight **Bar1** as in the screen capture above and then click on **Set Parameters** to receive the following dialog box.

Make sure that **Au̲tomatic** is selected as the option in the first box. In the second box, select **Cus̲tom** and set the **Number of intervals** at 18 (the number of different IQ scores received by the 30 participants in the study). Then click **Continue** and then **Apply**. Finally, click **OK** in the dialog box on the left, and you should receive the histogram in the output file.

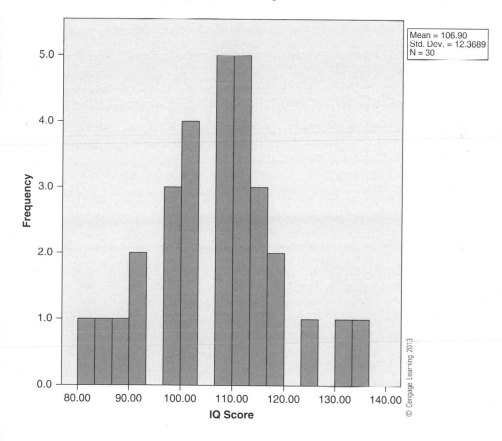

Notice that the bars are touching, except for those instances in which there were missing scores.

Frequency Polygons

We can also depict the data in a histogram as a *frequency polygon*—a line graph of the frequencies of individual scores or intervals. Again, scores (or intervals) are shown on the x-axis and frequencies on the y-axis. Once all the frequencies are plotted, the data points are connected. We can use the data from Table 2.2 to construct a frequency polygon. Frequency polygons are appropriate when the variable is quantitative or the data are ordinal, interval, or ratio. In this respect, frequency polygons are similar to histograms. Frequency polygons are especially useful for continuous data (such as age, weight, or time) in which it is theoretically possible for values to fall anywhere along the continuum. For example, an individual can weigh 120.5 pounds or be 35.5 years of age. Histograms are more appropriate when the data are discrete (measured in whole units)—for example, the number of college classes taken or the number of siblings.

Using Excel to Create a Frequency Polygon (Line Graph)

To create a frequency polygon for Excel, begin by entering the data into an Excel spreadsheet, as follows. Then highlight only the Frequency data as is illustrated in the next screen capture.

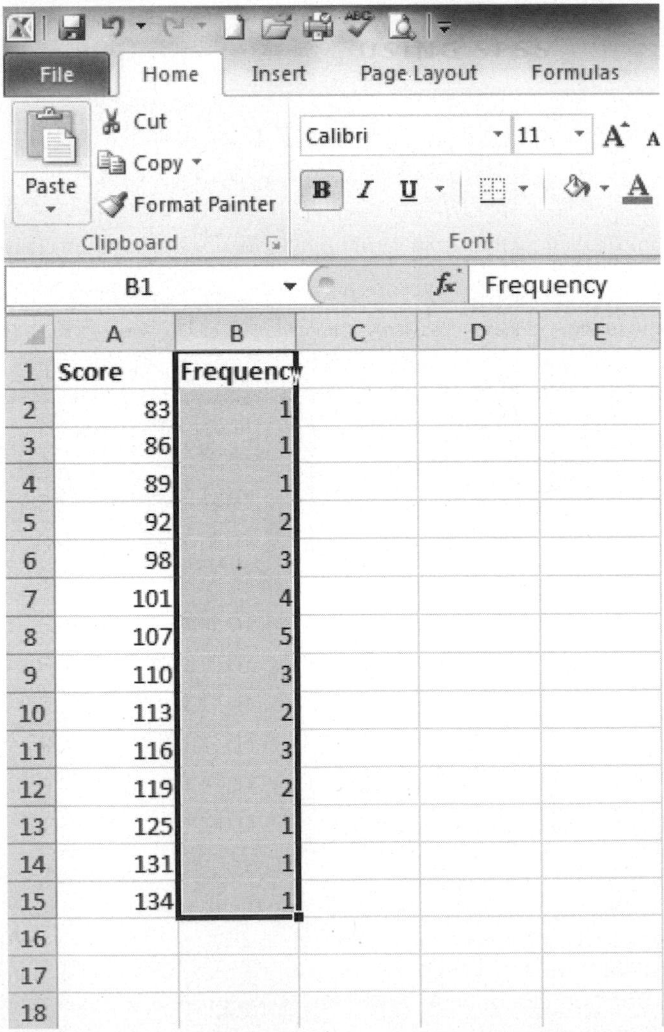

Next, click on the **Insert** ribbon and then **Line**. Select the option at the top left corner (the first 2-D line option). This should produce the following graph.

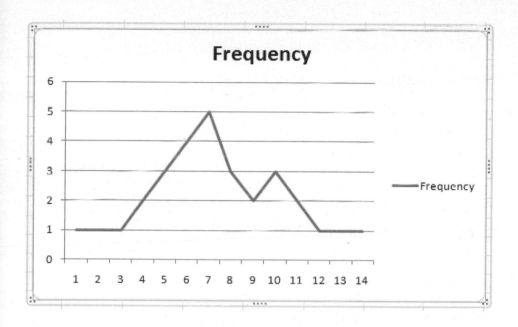

Next, right click on a value on the x-axis and then click **Select Data** to produce the following pop-up window:

Click on the **Edit** window under **Horizontal (Category) Axis Labels**. You'll receive the following pop-up window.

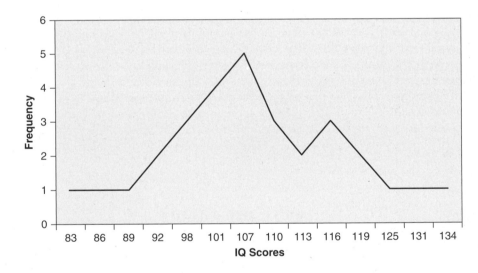

Highlight the IQ scores from the spreadsheet, and they will be inserted into the **Axis label range:** box. Then click **OK**. Click **OK** a second time to close the original pop-up window. You can now use the **Chart Tools** to format your frequency polygon so that it more closely resembles line graphs appropriate for APA style. This procedure will include adding axis labels to the x- and y-axes, changing the line from blue to black, and removing the gridlines from the graph. After making these changes, your figure should look as follows.

Using SPSS to Create a Frequency Polygon (Line Graph)

We'll once again use the data from Table 2.2 to illustrate how to create a frequency polygon using SPSS. Enter the data in the same manner as when we created a histogram in SPSS. In other words, enter each individual score on a separate line in SPSS so that all 30 scores in the distribution are entered individually. Once the data are entered, named, and coded as numeric with the scale level of measurement, click on **Graphs** and then **Chart Builder** to receive the following dialog boxes.

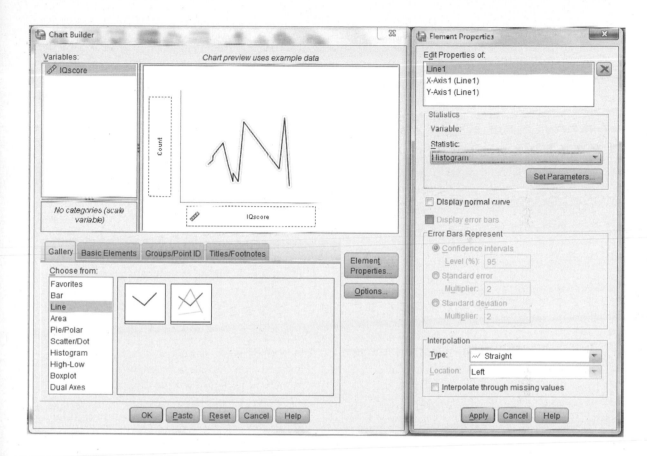

Double click on the **Line** graph option in the lower left of the screen and then drag the **IQscore** variable from the top left of the screen to *x*-axis box. In the **Element Properties** dialog box at the right of the screen, highlight **Line 1** and then **Histogram** in the **Statistic** box. Click on the **Set Parameters** box to receive the following dialog box.

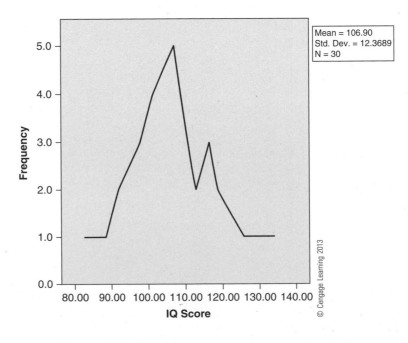

Select **A̲utomatic** in the first box, and then **Cus̲tom** in the second box indicating that the number of intervals should be 52 (the total range of IQ scores for our group of 30 individuals). Click **Continue** and then **Apply**. Finally, click **OK** to execute the procedure. You should receive the following frequency polygon.

© Cengage Learning 2013

TABLES

Using SPSS to Create a Table

To illustrate how to create a table with SPSS, let's use the data set presented in Table 2.3. These data represent the letter grade of 30 students on an introductory psychology exam and should be entered into SPSS exactly as they appear in

T A B L E 2.3   Letter Grade on Exam for 30 Students

© Cengage Learning 2012

Letter Grade	Gender
F	F
F	M
F	M
F	F
F	M
D	F
D	F
D	F
D	M
D	M
C	F
C	M
C	F
C	M
C	F
C	F
C	M
C	M
C	M
B	F
B	F
B	M
B	F
B	M
B	M
A	F
A	M
A	F
A	F
A	M

Table 2.3. Each student's gender has also been indicated in the table. One reason for organizing data and using statistics is so that meaningful conclusions can be drawn. As you can see from Table 2.3, our list of exam scores is simply that—a list. As shown here, the data are not especially meaningful. One of the first steps in organizing these data might be to condense the data into a *frequency distribution*, that is, a table in which all of the scores are listed along with the frequency with which each occurs. We can add further to the information by listing the frequency by each gender. The frequency distribution is a way of presenting data that makes the pattern of the data easier to see.

After entering the data into SPSS as they appear in Table 2.3, click on the **Analyze** menu, followed by **Tables**, and then **Custom Tables** to receive the following dialog box. Please note that this procedure can only be completed if you have the Custom Tables add-on module.

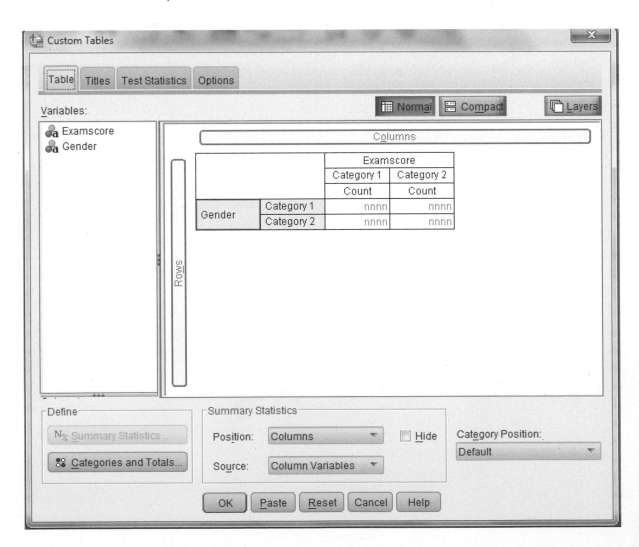

Select the **Examscore** variable and drag it to the **Columns** heading. Then select the **Gender** variable and drag it to the **Rows** heading. Then click **OK**. You will receive the custom table below that shows the frequency of each letter grade by gender.

		Examscore				
		A	B	C	D	F
		Count	Count	Count	Count	Count
Gender	F	3	3	4	3	2
	M	2	3	5	2	3

MODULE EXERCISES

(Answers appear in Appendix.)

The following data represent a distribution of speeds at which individuals were traveling on a highway:

64	80	64	70
76	79	67	72
65	73	68	65
67	65	70	62
67	68	65	64

1. Use either Excel or SPSS to create a histogram of these data.
2. Use either Excel or SPSS to create a frequency polygon of these data.
3. Use either Excel or SPSS to create a bar graph for the following data representing the number of freshmen, sophomores, juniors, and seniors enrolled in a particular class:

Freshmen	10
Sophomores	8
Juniors	7
Seniors	5

CHAPTER 2

✳

Descriptive Statistics

In this chapter, we will learn how to use Excel, SPSS, and the TI-84 calculator to determine various descriptive statistics. *Descriptive statistics* are numerical measures that describe a distribution by providing information on the central tendency of the distribution, the width of the distribution, and the shape of the distribution. We'll begin by using these tools to calculate a measure of central tendency—the mean. A *measure of central tendency* characterizes an entire set of data in terms of a single representative number. Measures of central tendency measure the "middleness" of a distribution of scores.

In addition, as part of the descriptive statistics procedures using these tools, we will also calculate a *measure of variability*—the standard deviation. We use measures of variability to assess the width of a distribution. A measure of variability indicates how scores are dispersed around the mean of the distribution.

Lastly, in Module 4, we'll use Excel to determine one additional descriptive statistic as z score or standard score. A *standard score* or z-*score* is a measure of how many standard deviation units an individual raw score falls from the mean of the distribution. We use such scores when we want information about a single score. For example, in a distribution of exam scores, we may want to know how one person's exam score compares with those of others in the class. Or we may want to know how an individual's exam score in one class, say psychology, compares with the same person's exam score in another class, say English. Because the two distributions of exam scores are different (different means and standard deviations), simply comparing the raw scores on the two exams will not provide this information.

Module 3

✳

Central Tendency
and Variability

The data we'll be using to illustrate how to calculate descriptive statistics are exam score data for a class of 30 students and are presented in the Table 3.1.

TABLE 3.1   Exam Scores Ordered from Lowest to Highest
© Cengage Learning 2013

Score	Score
45	76
47	77
54	78
56	78
59	80
60	82
60	82
63	85
65	86
69	87
70	90
74	92
74	93
74	94
75	95

USING EXCEL

To begin using Excel to conduct data analyses, the data must be entered into an Excel spreadsheet. Simply open Excel and enter the data into the spreadsheet. You can see in the following Excel spreadsheet that I have entered the exam grade data from Table 3.1.

	A	B	C	D	E	F
1	**Score**					
2	45					
3	47					
4	54					
5	56					
6	59					
7	60					
8	60					
9	63					
10	65					
11	69					
12	70					
13	74					
14	74					
15	74					
16	75					
17	76					
18	77					
19	78					
20	78					
21	80					
22	82					
23	82					
24	85					
25	86					
26	87					
27	90					
28	92					
29	93					
30	94					
31	95					

Once the data have been entered, we use the Data Analysis tool to calculate descriptive statistics. Click on the **Data** tab or ribbon and then click the **Data Analysis** icon at the far top right side of the window. Once the **Data Analysis** tab is active, a dialog box of options will appear (see next).

Select **Descriptive Statistics** as is indicated in the preceding box and then click **OK**. This will lead to the following dialog box.

With the cursor in the **Input Range** box, highlight the data that you want analyzed from column A in the Excel spreadsheet so that they appear in the input range. In addition, check the **Summary statistics** box. When you have done so, click **OK**. The summary statistics will appear in a new worksheet, as seen in the next screen capture.

	A	B
1	*Column1*	
2		
3	Mean	74
4	Standard Error	2.532546762
5	Median	75.5
6	Mode	74
7	Standard Deviation	13.8713299
8	Sample Variance	192.4137931
9	Kurtosis	-0.60744581
10	Skewness	-0.391850234
11	Range	50
12	Minimum	45
13	Maximum	95
14	Sum	2220
15	Count	30

As you can see, there are several descriptive statistics reported, including measures of central tendency (mean, median, and mode) and measures of variation (range, standard deviation, and sample variance).

USING SPSS

For SPSS, as with the Excel exercise above, we will once again be using the data from Table 3.1 to calculate descriptive statistics. We begin by entering the data from Table 3.1 into an SPSS spreadsheet. Simply open SPSS and enter the data into the spreadsheet. You can see in the following SPSS spreadsheet that I have entered the exam grade data from Table 3.1.

File Edit View Data Transform Analyze Graphs Utilities Add-ors Window Help

	VAR00001	var	var	var	var	var	var	var	var	var	var
1	45.00										
2	47.00										
3	54.00										
4	56.00										
5	59.00										
6	60.00										
7	60.00										
8	63.00										
9	65.00										
10	69.00										
11	70.00										
12	74.00										
13	74.00										
14	74.00										
15	75.00										
16	76.00										
17	77.00										
18	78.00										
19	78.00										
20	80.00										
21	82.00										
22	82.00										
23	85.00										
24	86.00										
25	87.00										
26	90.00										
27	92.00										
28	93.00										
29	94.00										
30	95.00										
31											

Notice that the variable is simply named VAR0001. To rename the variable to something appropriate for your data set, click on the **Variable View** tab at the bottom left of the screen. You will see the following window.

	Name	Type	Width	Decimals	Label	Values	Missing	Columns	Align	Measure	Role
1	VAR00001	Numeric	8	2		None	None	8	☰ Right	✐ Scale	↘ Input
2											
3											
4											
5											
6											
7											
8											
9											
10											
11											
12											
13											
14											
15											
16											
17											
18											
19											
20											
21											
22											
23											
24											
25											
26											
27											
28											
29											
30											
31											
32											
33											
34											
35											
36											

Data View   Variable View

Type the name you wish to give the variable in the highlighted **Name** box. The variable name cannot have any spaces in it. Because these data represent exam grade data, we'll type in **Examgrade**. Note also that the **Type** of data is Numeric. Once the variable is named, highlight the **Data View** tab at the bottom left of the screen in order to get back to the data spreadsheet. Once you've navigated back to the data spreadsheet, click the **Analyze** tab at the top of the screen and a drop-down menu with various statistical analyses will appear. Select **Descriptive Statistics** and then **Descriptives**. The following dialog box will appear.

Descriptives

Examgrade

Variable(s):

Options...

☐ Save standardized values as variables

OK   Paste   Reset   Cancel   Help

Examgrade will be highlighted, as above. Click on the arrow in the middle of the window and the Examgrade variable will be moved over to the **Variable(s)** box. Then click on **Options** to receive the following dialog box.

Descriptives: Options

☑ Mean   ☐ Sum

Dispersion

☑ Std. deviation   ☑ Minimum

☐ Variance   ☑ Maximum

☐ Range   ☐ S.E. mean

Distribution

☐ Kurtosis   ☐ Skewness

Display Order

◉ Variable list

○ Alphabetic

○ Ascending means

○ Descending means

Continue   Cancel   Help

You can see that the Mean, Standard Deviation, Minimum, and Maximum are all checked. However, you could select any of the descriptive statistics you want calculated. After making your selections, click **Continue** and then **OK**. The output will appear on a separate page as an output file like the one below in which you can see the minimum and maximum scores for this distribution along with the mean exam score of 74 and the standard deviation of 13.87. Please note that if you had more than one set of data, for example two classes of exam scores, they could each occupy one column in your SPSS spreadsheet, and you could conduct analyses on both variables at the same time. In such a situation, separate descriptive statistics would be calculated for each data set.

Descriptives

Descriptive Statistics

	N	Minimum	Maximum	Mean	Std. Deviation
Exam Grade	30	45.00	95.00	74.0000	13.87133
Valid N (listwise)	30				

USING THE TI-84

Follow the steps below to use your TI-84 calculator to calculate the mean for the data set from Table 3.1.

TI-84 Exercise: Calculation of the Mean:

1. With the calculator on, press the **STAT** key.
2. **EDIT** will be highlighted. Press the **ENTER** key.
3. Under L1 enter the data from Table 3.1.
4. Press the **STAT** key again and highlight **CALC**.
5. Number **1: 1—VAR STATS** will be highlighted. Press **ENTER**.
6. Press **ENTER** once again.

The statistics for the single variable on which you entered data will be presented on the calculator screen. The mean is presented on the first line of output as $\overline{X}$.

Follow the steps below to use your TI-84 calculator to calculate the standard deviation for the data set from Table 3.1.

TI-84 Exercise: Calculation of σ (standard deviation for population) and s (estimated population standard deviation):

1. With the calculator on, press the **STAT** key.
2. **EDIT** will be highlighted. Press the **ENTER** key.
3. Under L1 enter the data from Table 3.1.
4. Press the **STAT** key once again and highlight **CALC**.

5. Number **1: 1—VAR STATS** will be highlighted. Press **ENTER**.

6. Press **ENTER** once again.

Descriptive statistics for the single variable on which you entered data will be shown. The population standard deviation (σ) is indicated by the symbol σ_X. The unbiased estimator of the population standard deviation (s) is indicated by the symbol S_X.

MODULE EXERCISES

(Answers appear in Appendix.)

Use either Excel, SPSS, or the TI-84 to calculate the mean for the following four distributions (Exercises 1–4).

1. 2, 2, 4, 5, 8, 9, 10, 11, 11, 11

2. 1, 2. 3. 4. 4. 5, 5, 5, 6, 6, 8, 9

3. 1, 3, 3, 3, 5, 5, 8, 8, 8, 9, 10, 11

4. 2, 3, 4, 5, 6, 6, 6, 7, 8, 8

Use either Excel, SPSS, or the TI-84 to calculate the standard deviation for the following five distributions (Exercises 5–9).

5. 1, 2, 3, 4, 5, 6, 7, 8, 9

6. −4, −3, −2, −1, 0, 1, 2, 3, 4

7. 10, 20, 30, 40, 50, 60, 70, 80, 90

8. 1, .2, .3, .4, .5, .6, .7, .8, .9

9. 100, 200, 300, 400, 500, 600, 700, 800, 900

Module 4

Standard Scores or *z* Scores

USING EXCEL TO DETERMINE *Z* SCORES

To illustrate how to calculate standard scores or *z* scores, we'll use the data from Table 4.1. We can see in the table the mean and standard deviation for two groups of students: one group who took an English exam and a second group who took a psychology exam.

TABLE 4.1 **Mean and Standard Deviation for English and Psychology Exams for Two Classes**
© Cengage Learning 2013

	Mean	Standard Deviation
English	85	9.58
Psychology	74	13.64

The scores for two of the students who took each of these exams follow:

Person	English Exam	Psychology Exam
Rich	91	88
Debbie	82	80

We'll use Excel to calculate the *z* scores for each of these individuals on the two exams. To calculate *z* scores using Excel, we use a function other than the Data Analysis ToolPak. Open Excel and click on the **Formulas** tab. You can see in the following Excel worksheet that this tab is highlighted.

We'll start with the English exam data from Table 4.1 to calculate Rich's z score. You can see from that table that the English exam had a mean of 85 and a standard deviation of 9.58. To calculate the first z score, click on the *fX* button at the far left side of the **Formulas** ribbon. You should receive the following dialog box.

Make sure that **Statistical** is selected in the **Or select a _category_** field and then scroll down and select **STANDARDIZE** as in the preceding window. Finally, click **OK** to receive the following dialog box. Enter Rich's English exam score into the **X** box and the mean and standard deviation where indicated. Then click **OK**.

Excel will give you the preceding output in which you can see the z score of $+0.6263$ in the column A1. This z score indicates that Rich scored 0.63 standard deviations above the mean on the English exam. If we want to compare this score to his performance on the psychology exam or to Debbie's performance on the English exam, we must calculate these z scores also. Accordingly, to calculate Rich's z score on the psychology exam, we use the same procedure as above. Use this procedure to calculate Rich's z score on the psychology exam and Debbie's z scores for both the English and psychology exams. You should receive the following results:

Rich's psychology exam z score $= +1.03$

Debbie's English exam z score $= -.31$

Debbie's psychology exam z score $= +.44$

Thus, we can see that although Rich's English exam score was higher than his psychology exam score (91 vs. 88), his z score on the psychology exam was larger than his z score on the English exam, indicating that he was 1.03 standard deviations above the mean on the psychology exam but only .63 standard deviation above the mean on the English exam. Debbie, on the other hand, had a negative z score on the English exam, indicating that she scored $-.31$ standard deviation below the class mean. However, her psychology z score was $+.44$ standard deviation above the mean of the class, even though her raw score on the psychology exam was lower than her raw score on the English exam (80 vs. 82).

MODULE EXERCISES

(Answers appear in Appendix.)

Use Excel to calculate z scores for the following individuals based on an exam distribution with a mean of 55 and a standard deviation of 6.

1. John who scored 63 on the exam
2. Ray who scored 45 on the exam
3. Betty who scored 58 on the exam

✳

Introduction to Inferential Statistics

In this chapter, we introduce inferential statistics, that is, procedures for drawing conclusions about a population based on data collected from a sample. We will address two different statistical tests: the z test and t test. After reading this chapter, you should understand the differences between the two tests, when to use each test, and how to use the appropriate statistical tool to conduct each test.

Module 5

✳

The z Test

The *z test* is a parametric statistical test that allows you to test the null hypothesis for a single sample when the population variance is known. This procedure allows us to compare a sample to a population in order to assess whether the sample differs significantly from the population. If the sample was drawn randomly from a certain population and we observe a difference between the sample and a broader population, we can then conclude that the population represented by the sample differs significantly from the comparison population.

Suppose a researcher wants to examine the relationship between the type of after-school program attended by a child and intelligence level. The researcher is interested in whether students who attend an after-school program that is academically oriented (with such subjects as math, writing, and computer use, for example) score higher on an intelligence test than students who do not attend such a program. The researcher forms a hypothesis, which might state that children in academic after-school programs will have higher IQ scores than children in the general population. Because most intelligence tests are standardized with a mean score (μ) of 100 and a standard deviation (σ) of 15, the students in the academic after-school program would have to score higher than 100 for the hypothesis to be supported.

Let's assume that we have actually collected IQ scores from 75 students enrolled in academic after-school programs. We want to determine whether the sample of children in these programs represents a population with a mean IQ greater than the mean IQ of the general population of children. Let's assume that our sample of 75 students has an average IQ score of 103.5. This score represents the sample mean ($\overline{X}$). Because we are testing to see whether children in academic after-school programs have a higher general IQ score, we are using a one-tailed hypothesis. The null and alternative hypotheses for a one-tailed test are:

H_0: $\mu_0 \leq \mu_1$, or $\mu_{\text{academic program}} \leq \mu_{\text{general population}}$

H_a: $\mu_0 > \mu_1$, or $\mu_{\text{academic program}} > \mu_{\text{general population}}$

Excel and SPSS do not have point and click methods for conducting a z test, but the test can be conducted using the TI-84 calculator.

USING THE TI-84

1. With the calculator on, press the **STAT** key.
2. Highlight **TESTS**.
3. **1: Z-Test** will be highlighted. Press **ENTER**.
4. Highlight **STATS**. Press **ENTER**.
5. Scroll down to μ_0: and enter the mean for the population (100).
6. Scroll down to σ: and enter the standard deviation for the population (15).
7. Scroll down to $\overline{X}$: and enter the mean for the sample (103.5).
8. Scroll down to **n:** and enter the sample size (75).
9. Lastly, scroll down to μ: and select the type of test (one-tailed), indicating that we expect the sample mean to be greater than the population mean (select $> \mu_0$). Press **ENTER**.
10. Highlight **CALCULATE** and press **ENTER**.

The z score of 2.02 should be displayed followed by the significance level of .02. If you would like to see where the z score falls on the normal distribution, repeat Steps 1 through 9, then highlight **DRAW**, and press **ENTER**.

A z test score of 2.02 can be used to test our hypothesis that the sample of children in the academic after-school program represents a population with a mean IQ greater than the mean IQ for the general population. To do so, we need to determine whether the probability is high or low that a sample mean as large as 103.5 would be chosen from this sampling distribution. In other words, is a sample mean IQ score of 103.5 far enough away from or different enough from the population mean of 100 for us to say that it represents a significant difference with an alpha level of .05 or less? We can see above that the alpha level, or significance level is $p = .02$. Thus, the alpha level is less than .05, and the mean IQ scores of children in the sample differs significantly from that of children in the general population. In other words, children in academic after-school programs score significantly higher on IQ tests than children in the general population. In APA style, it would be reported as follows: $z \ (N = 75) = 2.02$, $p < .05$ (one-tailed).

TWO-TAILED Z TEST

The previous example illustrated a one-tailed z test. However, some hypotheses are two-tailed, and thus, the z test would also be two-tailed. As an example, imagine I am conducting a study to see whether children in athletic after-school programs weigh a different amount than children in the general population. In other words, we expect the weight of the children in the athletic after-school program to differ from that of children in the general population, but we are not sure whether they will weigh less (because of the activity) or more (because of greater muscle mass). H_0 and H_a for this two-tailed test are:

H_0: $\mu_0 = \mu_1$, or $\mu_{\text{athletic programs}} = \mu_{\text{general population}}$

H_a: $\mu_0 \neq \mu_1$, or $\mu_{\text{athletic programs}} \neq \mu_{\text{general population}}$

Let's use the following data: The mean weight of children in the general population (μ) is 90 pounds, with a standard deviation (σ) of 17 pounds; for children in the sample ($N=50$), the mean weight ($\overline{X}$) is 86 pounds. Using this information, we can now test the hypothesis that children in athletic after-school programs differ in weight from those in the general population using the TI-84 calculator.

Using the TI-84

1. With the calculator on, press the **STAT** key.
2. Highlight **TESTS**.
3. **1: Z-Test** will be highlighted. Press **ENTER**.
4. Highlight **STATS**. Press **ENTER**.
5. Scroll down to μ_0**:** and enter the mean for the population (90).
6. Scroll down to σ**:** and enter the standard deviation for the population (17).
7. Scroll down to $\overline{X}$**:** and enter the mean for the sample (86).
8. Scroll down to **n:** and enter the sample size (50).
9. Lastly, scroll down to μ**:** and select the type of test (two-tailed), indicating that we expect the sample mean to differ from the population mean (select $\neq \mu_0$). Press **ENTER**.
10. Highlight **CALCULATE** and press **ENTER**.

A z score of -1.66 should be displayed followed by the alpha level of .096, indicating that this test was not significant. We can therefore conclude that the weight of children in the athletic after-school program does not differ significantly from the weight of children in the general population.

If you would like to see where the z score falls on the normal distribution, repeat Steps 1 through 9, then highlight **DRAW**, and press **ENTER**.

MODULE EXERCISES

(Answers appear in Appendix.)

Use the TI-84 calculator to conduct the following z tests.

1. A researcher is interested in whether students who attend private high schools have higher average SAT scores than students in the general population of high school students. A random sample of 90 students at a private high school is tested and has a mean SAT score of 1050. The average for public high school students is 1000 ($\sigma = 200$).

    a. Is this a one- or two-tailed test?

    b. Compute the z test score.

    c. Draw conclusions from the test.

2. The producers of a new toothpaste claim that it prevents more cavities than other brands of toothpaste. A random sample of 60 people use the new toothpaste for 6 months. The mean number of cavities at their next checkup is 1.5. In the general population, the mean (μ) number of cavities at a 6-month checkup is 1.73 ($\sigma = 1.12$).

    a. Is this a one- or two-tailed test?

    b. Compute the z test score.

    c. Draw conclusions from the test.

Module 6

✳

The Single-Sample *t* Test

The *t test* for a single sample is similar to the z test in that it is also a parametric statistical test of the null hypothesis for a single sample. As such, it is a means of determining the number of standard deviation units a score is from the mean (μ) of a distribution. With a t test, however, the population variance is not known. Another difference is that t distributions, although symmetrical and bell-shaped, are *not* normally distributed. This means that the areas under the normal curve that apply for the z test do not apply for the t test.

Let's illustrate the use of the single-sample t test to test a hypothesis. Assume the mean SAT score of students admitted to General University is 1090. Thus, the university mean of 1090 is the population mean (μ). The population standard deviation is unknown. The members of the biology department believe that students who decide to major in biology have higher SAT scores than the general population of students at the university. The null and alternative hypotheses are thus:

H_0: $\mu_0 \leq \mu_1$, or $\mu_{\text{biology students}} \leq \mu_{\text{general population}}$

H_a: $\mu_0 > \mu_1$, or $\mu_{\text{biology students}} > \mu_{\text{general population}}$

Notice that this test is one-tailed because the researchers predict that the biology students will perform higher than the general population of students at the university. The researchers now need to obtain the SAT scores for a sample of biology majors. This information is provided in Table 6.1, which shows that the mean SAT score for a sample of 10 biology majors is 1176.

TABLE 6.1   **SAT Scores for a Sample of 10 Biology Majors**
© Cengage Learning 2013

1010
1200
1310
1075
1149
1078
1129
1069
1350
1390
$\bar{X} = 1176$

USING EXCEL

To demonstrate how to use Excel to calculate a single-sample *t* test, we'll use the data from Table 6.1. We are testing whether biology majors have higher average SAT scores than the population of students at General University. We begin by entering the data into Excel. We enter the sample data into Column A and the population mean of 1090 into Column B. We enter the population mean next to the score for each individual in the sample. You can see that I've entered 1090 in Column B ten times, one time for each individual in our sample of 10 biology majors.

Next, highlight the **Data** ribbon, and then click on **Data Analysis** at the top right-hand corner. You should now have the following pop-up window.

Scroll down to **t-Test: Paired Two Sample for Means**, which is the procedure we'll be using to determine the single-sample *t* test. Then click **OK**. You'll be presented with the following dialog box.

With the cursor in the **Variable 1 Range:** box, highlight the data from Column A in the Excel spreadsheet so that they appear in the input range box. Move the cursor to the **Variable 2 Range:** box and enter the data from Column B in the spreadsheet into this box by highlighting the data. The dialog box should now appear as follows.

t-Test: Paired Two Sample for Means

Input
Variable 1 Range: $A$2:$A$11
Variable 2 Range: $D$2:$B$11

Hypothesized Mean Difference:

☐ Labels
Alpha: 0.05

Output options
○ Output Range:
◉ New Worksheet Ply:
○ New Workbook

OK
Cancel
Help

Click **OK** to execute the problem. You will be presented with the following output.

	A	B	C	D	E
1	t-Test: Paired Two Sample for Means				
2					
3		Variable 1	Variable 2		
4	Mean	1176	1090		
5	Variance	17372.44444	0		
6	Observations	10	10		
7	Pearson Correlation	#DIV/0!			
8	Hypothesized Mean	0			
9	df	9			
10	t Stat	2.06332664			
11	P(T<=t) one-tail	0.034553462			
12	t Critical one-tail	1.833112933			
13	P(T<=t) two-tail	0.069106925			
14	t Critical two-tail	2.262157163			
15					

We see a *t* test score of 2.063 and a one-tailed significance level of $p = .035$. Therefore, our sample mean falls 2.06 standard deviations above the population mean of 1090. We must now determine whether this is far enough away from the population mean to be considered significantly different. Because our obtained alpha level (significance level) is .035 and is less than .05, the result is significant. We reject H_0 and support H_a. In other words, we have sufficient evidence to allow us to conclude that biology majors have significantly higher SAT scores than the rest of the students at General University. In APA style, this information would be reported as $t(9) = 2.06$, $p = .035$ (one-tailed).

USING SPSS

To demonstrate how to use SPSS to calculate a single-sample *t* test, we'll again use the data from Table 6.1. As before, we are testing whether biology majors have higher average SAT scores than the population of students at General University. We begin by entering the data into SPSS and naming the variable, as follows (if you've forgotten how to name a variable, please refer back to Module 1).

File	Edit	View	Data	Transform	Analyze	Graphs	Utilities	Add-ons	Window	Help

11 : SATscore

	SATscore	var	var	var	var	var
1	1010.00					
2	1200.00					
3	1310.00					
4	1075.00					
5	1149.00					
6	1078.00					
7	1129.00					
8	1069.00					
9	1350.00					
10	1390.00					

Then select the **Analyze** tab and, from the drop-down menu, **Compare Means** followed by **One-Sample T Test**. The following dialog box will appear.

Place the SATscore variable into the **Test Variable(s)** box by utilizing the arrow in the middle of the window. Then let SPSS know what the population mean SAT score is. We can find this earlier in this module where the problem is described: it is 1090. We enter this population mean in the **Test Value** box as in the following window.

Then click **OK,** and the output for the single–sample *t* test will be produced in an output window as follows.

T-Test

One-Sample Statistics

	N	Mean	Std. Deviation	Std. Error Mean
SATscore	10	1176.0000	131.80457	41.68026

One-Sample Test

	Test Value = 1090					
					95% Confidence Interval of the Difference	
	t	df	Sig. (2-tailed)	Mean Difference	Lower	Upper
SATscore	2.063	9	.069	86.00000	-8.2873	180.2873

We can see the t test score of 2.063 and the two–tailed significance level. Thus, our sample mean falls 2.06 standard deviations above the population mean of 1090. We must now determine whether this is far enough away from the population mean to be considered significantly different. This was a one-tailed test; accordingly, when using SPSS, you will need to divide the significance level in half to obtain a one-tailed significance level because SPSS reports only two-tailed significance levels. Our alpha level (significance level) is .035 and is less than .05, meaning the result is significant. We reject H_0 and support H_a. In other words, we have sufficient evidence to allow us to conclude that biology majors have significantly higher SAT scores than the rest of the students at General University. In APA style, this would be reported as $t(9) = 2.06$, $p = .035$ (one-tailed).

USING THE TI-84

Let's use the data from Table 6.1 to conduct the test using the TI-84 calculator.

1. With the calculator on, press the **STAT** key.
2. **EDIT** will be highlighted. Press the **ENTER** key.
3. Under **L1** enter the **SAT** data from Table 6.1.
4. Press the **STAT** key once again and highlight **TESTS**.
5. Scroll down to **T-Test**. Press the **ENTER** key.
6. Highlight **DATA** and press **ENTER**. Enter 1090 (the mean for the population) next to μ_0:. Enter L_1 next to **List** (to do this press the 2nd key followed by the **1** key).
7. Scroll down to μ: and select $> \mu_0$ (for a one-tailed test in which we predict that the sample mean will be greater than the population mean). Press **ENTER**.
8. Scroll down to and highlight **CALCULATE**. Press **ENTER**.

The t score of 2.06 should be displayed followed by the significance level of .035. In addition, descriptive statistics will be shown. If you would like to see where the t score falls on the distribution, repeat Steps 1 through 7, then highlight **DRAW**, and press **ENTER**.

MODULE EXERCISES

(Answers appear in Appendix.)

Use either SPSS or the TI-84 to calculate the t tests for the following problems.

1. A researcher hypothesizes that people who listen to music via headphones have greater hearing loss and will thus score lower on a hearing test than those in the general population. On a standard hearing test, $\mu = 22.5$. The researcher gives this same test to a random sample of 12 individuals who regularly use headphones. Their scores on the test are 16, 14, 20, 12, 25, 22, 23, 19, 17, 17, 21, 20.

    a. Is this a one- or two-tailed test?

    b. Compute the t test score.

    c. What should the researcher conclude?

2. A researcher hypothesizes that individuals who listen to classical music will score differently from the general population on a test of spatial ability. On a standardized test of spatial ability, $\mu = 58$. A random sample of 14 individuals who listen to classical music is given the same test. Their scores on the test are 52, 59, 63, 65, 58, 55, 62, 63, 53, 59, 57, 61, 60, 59.

    a. Is this a one- or two-tailed test?

    b. Compute the t test score.

    c. What should the researcher conclude?

CHAPTER 4

✳

t Tests for Two-group Designs

In this chapter, we discuss the common types of statistical analyses used with simple two-group designs. The inferential statistics discussed in this chapter differ from those presented in Chapter 3 in which single samples were being compared to populations (*z* test and *t* test). In this section, the statistics are designed to test differences between two equivalent groups of participants.

Several factors influence which statistic should be used to analyze the data collected. For example, the type of data collected and the number of groups being compared must be considered. Moreover, the statistic used to analyze the data will vary depending on whether the study involves a *between-subjects design* in which different participants are used in each of the groups or a *correlated-groups design* in which the participants in the experimental and control groups are related in some way. We will look at the typical inferential statistics used to analyze interval-ratio data for two-group between-subjects designs and correlated-groups designs.

In the two-group design, two samples (representing two populations) are compared by having one group receive nothing (the control group) and the second group receive some level of the manipulated variable (the experimental group). It is also possible to have two experimental groups and no control group. In this case, members of each group receive a different level of the manipulated variable.

Module 7

✳

Independent-Groups *t* Test

The *independent-groups t test* is a parametric statistical test that compares the performance of two different samples of participants. It indicates whether the two samples perform so similarly that we conclude that they are likely from the same population or whether they perform so differently that we conclude that they represent two different populations. Imagine, for example, that a researcher wants to study the effects on exam performance of massed versus spaced study. All participants in the experiment study the same material for the same amount of time. The difference between the groups is that one group studies for 6 hours all at once (massed study), whereas the other group studies for 6 hours broken into three 2-hour blocks (spaced study). Because the researcher believes that the spaced-study method will lead to better performance, the null and alternative hypotheses are:

H_0: Spaced Study $\leq$ Massed Study, or $\mu_1 \leq \mu_2$

H_a: Spaced Study $>$ Massed Study, or $\mu_1 > \mu_2$

The 20 participants are chosen by random sampling and randomly assigned to the groups. Because of the random assignment of participants, we are confident that there are no major differences between the groups prior to the study. The dependent variable is the participants' scores on a 30-item test of the material; these scores are listed in Table 7.1.

TABLE 7.1   Number of Items Answered Correctly by Each Participant Under Spaced - versus Massed-Study Conditions Using a Between-Subjects Design (*N* = 20)
© Cengage Learning 2013

Spaced Study	Massed Study
23	17
18	18
23	21
22	15
20	15
24	16
21	17
24	19
21	14
24	17
$\overline{X}_1 = 22$	$\overline{X}_2 = 16.9$

USING EXCEL

We'll use the data from Table 7.1 to illustrate how to use Excel to calculate an independent-groups *t* test. The data represent the number of items answered correctly for two groups of participants when one group used a spaced-study technique and the other used a massed-study technique. The researcher predicted that those in the spaced-study condition would perform better. The data from Table 7.1 have been entered into the following Excel worksheet with the data from the spaced condition in Column A and the data from the massed condition in Column B.

Next, with the **Data Ribbon** active, we click on **Data Analysis** in the top right corner of the screen and the following dialog box appears.

You can see that I have selected **t-test: Two-Sample Assuming Equal Variances**. After you have done the same, click **OK**, and you will then see the following dialog box.

t-Test: Two-Sample Assuming Equal Variances

Input

Variable 1 Range: $A$2:$A$11

Variable 2 Range: $B$2:$B$11

Hypothesized Mean Difference:

☐ Labels

Alpha: 0.05

Output options

○ Output Range:

◉ New Worksheet Ply:

○ New Workbook

OK   Cancel   Help

With the cursor in the **Variable 1 Range** box, highlight the data in Column A in the Excel spreadsheet so that they are entered into the **Variable 1 Range** box (do not highlight the column heading of **Spaced Study**). Do the same for Column B and enter these data into the **Variable 2 Range** box. Then click **OK**. You will see the following output.

File   Home   Insert   Page Layout   Formulas   Data   Review

From Access From Web From Text From Other Sources Existing Connections Refresh All Connections Properties Edit Links

Get External Data   Connections

F17   fx

	A	B	C	D
1	t-Test: Two-Sample Assuming Equal Variances			
2				
3		*Variable 1*	*Variable 2*	
4	Mean	22	16.9	
5	Variance	4	4.322222222	
6	Observations	10	10	
7	Pooled Variance	4.161111111		
8	Hypothesized Mea	0		
9	df	18		
10	t Stat	5.590498329		
11	P(T<=t) one-tail	1.31745E-05		
12	t Critical one-tail	1.734063607		
13	P(T<=t) two-tail	2.6349E-05		
14	t Critical two-tail	2.10092204		
15				

We are provided with the *t* test statistic of 5.59 along with the probability and critical values for both one- and two-tailed tests. We can see based on the one-tailed critical value of *t* that the test is significant at $p < .0000132$. This would be reported in APA style as $t (18) = 5.59, p = .0000132$ (one-tailed).

USING SPSS

We'll employ the same problem to illustrate the use of SPSS for an independent-groups *t* test. As above, researchers have participants use one of two types of study, spaced or massed, and then they measured exam performance. The data from Table 7.1 are entered into SPSS as in the following window.

File	Edit	View	Data	Transform	Analyze	Graphs	Utilities	Add-ons	Window	Help

1 : Typeofstudy		1.0				
	Typeofstudy	Examscore	var	var	var	var
1	1.00	23.00				
2	1.00	18.00				
3	1.00	23.00				
4	1.00	22.00				
5	1.00	20.00				
6	1.00	24.00				
7	1.00	21.00				
8	1.00	24.00				
9	1.00	21.00				
10	1.00	24.00				
11	2.00	17.00				
12	2.00	18.00				
13	2.00	21.00				
14	2.00	15.00				
15	2.00	15.00				
16	2.00	16.00				
17	2.00	17.00				
18	2.00	19.00				
19	2.00	14.00				
20	2.00	17.00				
21						

Notice that the independent variable of **Type of Study** has been converted to a numeric variable where the number 1 represents the spaced-study condition and the number 2 represents the massed-study condition. Thus, the data in rows 1 through 10 represent spaced-study data, and the data in rows 11 through 20 represent the massed–study data. Click on the **Analyze** tab and then **Compare Means** followed by **Independent–Samples T Test** as is illustrated next.

	Typeofstudy	Exams				ar	var	var	var
1	1.00	2							
2	1.00	1							
3	1.00	2							
4	1.00	2							
5	1.00	2							
6	1.00	2							
7	1.00	2							
8	1.00	2							
9	1.00	2							
10	1.00	2							
11	2.00	17.00							
12	2.00	18.00							
13	2.00	21.00							
14	2.00	15.00							
15	2.00	15.00							
16	2.00	16.00							
17	2.00	17.00							
18	2.00	19.00							
19	2.00	14.00							
20	2.00	17.00							
21									

Analyze menu:
Reports
Descriptive Statistics
Compare Means → Means...
One-Sample T Test...
Independent-Samples T Test...
Paired-Samples T Test...
One-Way ANOVA...
General Linear Model
Correlate
Regression
Classify
Dimension Reduction
Scale
Nonparametric Tests
Forecasting
Multiple Response
Quality Control
ROC Curve...

The following dialog box will appear.

Independent-Samples T Test

Typeofstudy
Examscore

Test Variable(s):

Options...

Grouping Variable:

Define Groups...

OK   Paste   Reset   Cancel   Help

We'll place the Exam score data into the **Test Variable** (dependent variable) box and the Type of study data into the **Grouping Variable** (independent variable) box by highlighting each variable and using the arrow keys in the middle of the dialog box to move the variables. The dialog box should appear as follows once you've completed this task.

Once you have completed this operation, click on the **Grouping Variable** box and the **Define Groups** box below it, which will become active, and you will receive a dialog box as follows.

We have to let SPSS know what values we used to designate the spaced versus the massed study groups. Thus, enter a 1 into the **Group 1** box and a 2 into the **Group 2** box and click **Continue**. Then click **OK** in the **Independent-Samples T Test** dialog box. You should receive output similar to the following.

T-Test

Group Statistics

	Typeofstudy	N	Mean	Std. Deviation	Std. Error Mean
Examscore	1.00	10	22.0000	2.00000	.63246
	2.00	10	16.9000	2.07900	.65744

Independent Samples Test

		Levene's Test for Equality of Variances		t-test for Equality of Means						95% Confidence Interval of the Difference	
		F	Sig.	t	df	Sig. (2-tailed)	Mean Difference	Std. Error Difference	Lower	Upper
Examscore	Equal variances assumed	.022	.884	5.590	18	.000	5.10000	.91226	3.18341	7.01659
	Equal variances not assumed			5.590	17.973	.000	5.10000	.91226	3.18320	7.01680

Descriptive statistics for the two conditions are reported in the first table followed by the *t* test score of 5.590. Because we are assuming equal variances, we use the *df*, *t* score, and other data from that row in the table. Moreover, the two-tailed significance level is provided, so we would have to divide this number in half to obtain the one-tailed significance level. However, because SPSS is reporting only the first three numbers, the significance level is reported as .000. Remember from the Excel example earlier that the significant level when carried out several decimal places is .0000132. We are also provided with the 95% confidence interval for the *t* test.

USING THE TI-84

Let's use the data from Table 7.1 to conduct the test using the TI-84 calculator:

1. With the calculator on, press the **STAT** key.
2. **EDIT** will be highlighted. Press the **ENTER** key.
3. Under **L1** enter the data from Table 7.1 for the spaced-study group.
4. Under **L2** enter the data from Table 7.1 for the massed-study group.
5. Press the **STAT** key once again and highlight **TESTS**.
6. Scroll down to **2-SampTTest**. Press the **ENTER** key.
7. Highlight **DATA**. Enter **L1** next to **List1** (by pressing the 2nd key followed by the **1** key). Enter **L2** next to **List2** (by pressing the 2nd key followed by the **2** key).

8. Scroll down to $\mu1$: and select $> \mu2$ (for a one-tailed test in which we predict that the spaced-study group will do better than the massed-study group). Press **ENTER**.

9. Scroll down to **Pooled:** and highlight **YES**. Press **ENTER**.

10. Scroll down to and highlight **CALCULATE**. Press **ENTER**.

The *t* score of 5.59 should be displayed followed by the significance level of .000013 and the *df* of 18. In addition, descriptive statistics for both variables on which you entered data will be shown.

MODULE EXERCISES

(Answers appear in Appendix.)

1. A college student is interested in whether there is a difference between male and female students in the amount of time spent studying each week. The student gathers information from a random sample of male and female students on campus. Amount of time spent studying is normally distributed. The data follow.

Males	Females
27	25
25	29
19	18
10	23
16	20
22	15
14	19

© Cengage Learning 2013

Conduct the independent-groups *t* test using Excel, SPSS, and the TI-84 calculator.

2. A student is interested in whether students who study with music playing devote as much attention to their studies as do students who study under quiet conditions. He randomly assigns the 18 participants to either music or no music conditions and has them read and study the same passage of information for the same amount of time. Participants are then all given the same 10-item test on the material. Their scores appear below. Scores on the test represent interval ratio data and are normally distributed.

Music	No Music
6	10
5	9
6	7
5	7
6	6
6	6
7	8
8	6
5	9

Conduct the independent-groups *t* test using Excel, SPSS, and the TI-84 calculator.

Module 8

✳

Correlated-Groups *t* Test

USING EXCEL

The *correlated-groups t test*, like the previously discussed independent-groups *t* test, compares the performance of participants in two groups. In this case, however, the same people are used in each group (a within-subjects design) or different participants are matched between groups (a matched-subjects design). The test indicates whether there is a difference in sample means and whether this difference is greater than would be expected based on chance. In a correlated-groups design, the sample includes two scores for each person (or matched pair in a matched-subjects design) instead of just one. The null hypothesis is that there is no difference between the two scores, that is, a person's score in one condition is the same as that (or a matched) person's score in the second condition. The alternative hypothesis is that there is a difference between the paired scores—that the individuals (or matched pairs) performed differently in each condition.

To illustrate the use of the correlated-groups *t* test, imagine that we conduct a study in which participants are asked to learn two lists of words. One list is composed of 20 concrete words (for example, *desk, lamp,* and *bus*), whereas the other is composed of 20 abstract words (for example, *love, hate,* and *deity*). Each participant is tested twice, once in each condition. Because each participant provides one pair of scores, a correlated-groups *t* test is the appropriate way to compare the means of the two conditions. We expect to find that recall performance is better for the concrete words. Thus, the null hypothesis is H_0: $\mu_{\text{Concrete}} \leq \mu_{\text{Abstract}}$, and the alternative hypothesis is H_a: $\mu_{\text{Concrete}} > \mu_{\text{Abstract}}$, representing a one-tailed test of the null hypothesis.

Using Excel to calculate a correlated-groups *t* test is very similar to using it to calculate an independent-groups *t* test. We'll use the data from Table 8.1 to illustrate its use.

TABLE 8.1   **Number of Abstract and Concrete Words Recalled by Each Participant Using a Correlated-Groups (Within-Subjects) Design**

© Cengage Learning 2013

Participant	Concrete	Abstract
1	13	10
2	11	9
3	19	13
4	13	12
5	15	11
6	10	8
7	12	10
8	13	13

For this *t* test, we are comparing memory for concrete versus abstract words for a group of 8 participants. Each participant served in both conditions. First enter the data from Table 8.1 into an Excel spreadsheet (as seen next). The data for the concrete-word condition are entered into Column A and the data for the abstract-word condition into Column B.

Then, with the **Data** ribbon active, click on **Data Analysis** and select **t–test: Paired Two Sample for Means** as indicated in the following dialog box. Click **OK** after completing this operation.

You will then get the following dialog box into which you will enter the data from Column A into the **Variable 1 Range** box by clicking in the **Variable 1 Range** box and then highlighting the data in Column A and then doing the same with the data in Column B and the **Variable 2 Range box**. The dialog box should appear as follows.

Click **OK**, and you will receive the output as it appears next.

⊿	A	B	C	D	E	F	G
1	t-Test: Paired Two Sample for Means						
2							
3		Variable 1	Variable 2				
4	Mean	13.25	10.75				
5	Variance	7.642857143	3.35714286				
6	Observations	8	8				
7	Pearson Correlation	0.747368952					
8	Hypothesized Mean Difference	0					
9	df	7					
10	t Stat	3.818813079					
11	P(T<=t) one-tail	0.003276179					
12	t Critical one-tail	1.894578605					
13	P(T<=t) two-tail	0.006552357					
14	t Critical two-tail	2.364624252					
15							
16							
17							

We can see that $t(7) = 3.82$, $p = .0033$ (one-tailed).

USING SPSS

To illustrate using SPSS for the correlated-groups *t* test, we'll use the same prob-lem described above in which a researcher has a group of participants study a list of 20 concrete words and 20 abstract words and then measures recall for the words within each condition. The researcher predicts that the participants will have better recall for the concrete words. The data from Table 8.1 are entered into SPSS as follows. We have 8 participants, and each serves in both conditions. Thus, the scores for each participant in both conditions appear in a single row.

File	Edit	View	Data	Transform	Analyze	Graphs	Utilities	Add-ons	Window	Help

| 1 : Participant | 1.0 |

	Participant	Concretewords	Abstractwords	var	var	var
1	1.00	13.00	10.00			
2	2.00	11.00	9.00			
3	3.00	19.00	13.00			
4	4.00	13.00	12.00			
5	5.00	15.00	11.00			
6	6.00	10.00	8.00			
7	7.00	12.00	10.00			
8	8.00	13.00	13.00			
9						
10						

Next, we click on the **Analyze** tab followed by the **Compare Means** tab and then **Paired–Samples T Test** as follows.

*Untitled1 [DataSet0] - PASW Statistics Data Editor

File	Edit	View	Data	Transform	Analyze	Graphs	Utilities	Add-ons	Window	Help

Reports ▶
Descriptive Statistics ▶
Compare Means ▶ → M Means...
General Linear Model ▶   t One-Sample T Test...
Correlate ▶   Independent-Samples T Test...
Regression ▶   Paired-Samples T Test...
Classify ▶   F One-Way ANOVA...
Dimension Reduction ▶
Scale ▶
Nonparametric Tests ▶
Forecasting ▶
Multiple Response ▶
Quality Control ▶
ROC Curve...

	Participant	Concr						var	var
1	1.00								
2	2.00								
3	3.00								
4	4.00								
5	5.00								
6	6.00								
7	7.00								
8	8.00								
9									
10									
11									
12									

These actions will produce the following dialog box.

Highlight the **Concretewords** variable and then click the arrow button in the middle of the screen. The Concretewords variable should now appear under **Variable1** in the box to the right of the window. Do the same for the **Abstractwords** variable, and it should appear under **Variable2** in the box on the right. The dialog box should now appear as follows.

Click **OK**, and the output will appear in an output window as below.

T-Test

Paired Samples Statistics

		Mean	N	Std. Deviation	Std. Error Mean
Pair 1	Concretewords	13.2500	8	2.76457	.97742
	Abstractwords	10.7500	8	1.83225	.64780

Paired Samples Correlations

		N	Correlation	Sig.
Pair 1	Concretewords & Abstractwords	8	.747	.033

Paired Samples Test

		Paired Differences					t	df	Sig. (2-tailed)
					95% Confidence Interval of the Difference				
		Mean	Std. Deviation	Std. Error Mean	Lower	Upper			
Pair 1	Concretewords - Abstractwords	2.50000	1.85164	.65465	.95199	4.04801	3.819	7	.007

As in the independent-samples t test in the previous module, descriptive statistics appear in the first table, followed by the correlation between the variables. The correlated-groups t test results appear in the third table with the t score of 3.819, 7 degrees of freedom, and the two-tailed significance level. Because this was a one-tailed test, we can find the significance level for this one-tailed t test by dividing the two-tailed significant level in half. Thus, for this problem t $(7) = 3.82$, $p = .0035$ (one-tailed). As in the previous t test in Module 7, the 95% confidence interval is also reported.

USING THE TI-84

Let's use the data from Table 8.1 to conduct the test using the TI-84 calculator:

1. With the calculator on, press the **STAT** key.
2. **EDIT** will be highlighted. Press the **ENTER** key.
3. Under **L1** enter the data for concrete words.
4. Under **L2** enter the data for abstract words.
5. Move the cursor so that **L3** is highlighted and then enter the following formula: L1 − L2. Press **ENTER** (to enter **L1** press the 2nd key followed by the **1** key and then to produce **L2**, press the 2nd key followed by the **2** key). This will produce a list of difference scores (the concrete-word scores minus the abstract-word scores) for each participant.
6. Press the **STAT** key once again and highlight **TESTS**.
7. Scroll down to **T-Test**. Press the **ENTER** key.

8. Highlight **DATA.** Enter 0 next to μ_0:. Enter **L3** next to **List** (by pressing the 2nd key followed by the **3** key).

9. Scroll down to μ: and select $> \mu_0$ (for a one-tailed test in which we predict that the difference between the scores for each condition will be greater than 0). Press **ENTER.**

10. Scroll down to and highlight **CALCULATE.** Press **ENTER.**

The *t* score of 3.82 should be displayed followed by the significance level of .0033. In addition, descriptive statistics will be shown.

MODULE EXERCISES

(Answers appear in Appendix.)

1. A researcher is interested in whether participating in sports positively influences self-esteem in young girls. She identifies a group of girls who have not played sports before but who are now planning to begin participating in organized sports. She gives them a 50-item self-esteem inventory before they begin playing sports and administers it again after 6 months of playing sports. The self-esteem inventory is measured on an interval scale, with higher numbers indicating higher self-esteem. In addition, scores on the inventory are normally distributed. The scores appear below.

Before	After
44	46
40	41
39	41
46	47
42	43
43	45

© Cengage Learning 2013

Conduct the correlated-groups *t* test using Excel, SPSS, and the TI-84 calculator.

2. The researcher in Question 2 from Module 7 decides to conduct the same study using a within-participants design in order to control for differences in cognitive ability. He selects a random sample of participants and has them study different material of equal difficulty in both the music and no music conditions. The data appear below. As before, they are measured on an interval-ratio scale and are normally distributed.

Music	No Music
6	10
7	7
6	8
5	7
6	7
8	9
8	8

© Cengage Learning 2013

Conduct the correlated-groups *t* test using Excel, SPSS, and the TI-84 calculator.

CHAPTER 5

✳

Analysis of Variance (ANOVA)

In this chapter, we discuss the common types of statistical analyses used with designs involving more than two groups. The inferential statistics discussed in this chapter differ from those presented in the previous two chapters in that in Chapter 3 single samples were being compared to populations (z test and t test) and in Chapter 4 two independent or correlated samples were being compared. In this chapter, the statistics are designed to test differences between more than two equivalent groups of participants.

Several factors influence which statistic should be used to analyze the data collected. For example, the type of data collected and the number of groups being compared must be considered. Moreover, the statistic used to analyze the data will vary depending on whether the study involves a between-subjects design (designs in which different participants are used in each group) or a correlated-groups design (one of two types, within-subjects designs in which the same subjects are used repeatedly in each group and matched-subjects designs in which different subjects are matched between conditions on variables that the researcher believes are relevant to the study.)

Module 9

✳

One-way Randomized ANOVA

For multiple-group designs in which interval-ratio data are collected, the recommended parametric statistical analysis is the *ANOVA (analysis of variance)*. As its name indicates, this procedure allows us to analyze the variance in a study. We will begin our coverage of statistics appropriate for multiple-group designs by discussing those used with data collected from a between-subjects design. Recall that a between-subjects design is one in which different participants serve in each condition. Imagine that we conducted a study in which participants are asked to study a list of 10 words using rote rehearsal or one of two forms of elaborative rehearsal (imagery or story). A total of 24 subjects are randomly assigned, 8 to each condition. Table 9.1 lists the number of words correctly recalled by each participant.

T A B L E 9.1   **Number of Words Recalled Correctly in Rote, Imagery, and Story Conditions**
© Cengage Learning 2013

Rote	Imagery	Story
2	4	6
4	5	5
3	7	9
5	6	10
2	5	8
7	4	7
6	8	10
3	5	9

Because these data represent an interval-ratio scale of measurement and because there are more than two groups, an ANOVA is the appropriate statistical test to analyze the data because the ANOVA is an inferential statistical test for comparing the means of three or more groups. In addition, because this is a between-subjects design, we use a *one-way randomized ANOVA*. The term *randomized* indicates that participants have been randomly assigned to conditions in a between-subjects design. The term *one-way* indicates that the design uses only one independent variable—in this case, type of rehearsal. We will discuss statistical tests appropriate for correlated-groups designs and tests appropriate for designs with more than one independent variable in Modules 10 and 11. Please note that although the study used to illustrate the ANOVA procedure in this section has an equal number of participants in each condition, this provision is not necessary to the procedure.

For the experiment described above, we are interested in the effects of rehearsal type on memory. The null hypothesis (H_0) for an ANOVA is that the sample means represent the same population (H_0: $\mu_1 = \mu_2 = \mu_3$). The alternative hypothesis (H_a) is that they represent different populations (H_a: at least one $\mu \neq$ another μ). When a researcher rejects H_0 using an ANOVA, it means that the independent variable affected the dependent variable to the extent that at least one group mean differs from the others by more than would be expected based on chance. Failing to reject H_0 indicates that the means do not differ from each other more than would be expected based on chance. In other words, there is not enough evidence to suggest that the sample means represent at least two different populations.

USING EXCEL

We'll use the data from Table 9.1 to illustrate the use of Excel to compute a one-way randomized ANOVA. In the study reported in the table, we had participants use one of three different types of rehearsal (rote, imagery, or story) and then had them perform a recall task. Consequently, we manipulated rehearsal and measured memory for the 10 words the participants studied. Because there were different participants in each condition, we use a randomized ANOVA.

We begin by entering the data into Excel, with the data from each condition appearing in a different column. This operation can be seen in the following image.

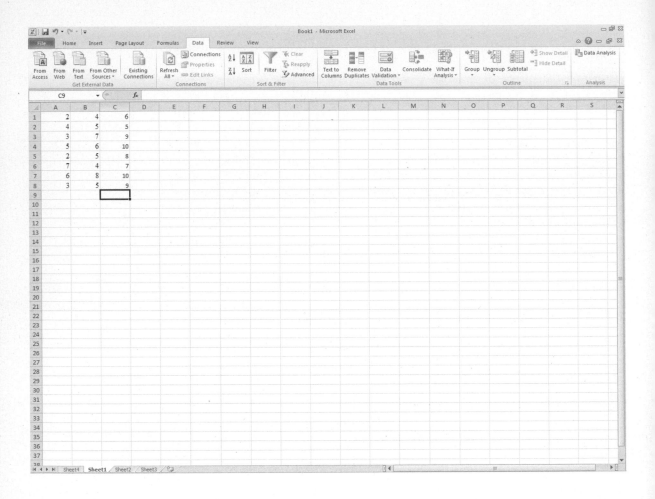

Next, with the **Data** ribbon highlighted, click on the **Data Analysis** tab at the top right corner. You should receive the following dialog box.

Select **Anova: Single Factor** in the preceding box and click **OK**. The following dialog box will appear.

With the cursor in the **Input Range** box, highlight the three columns of data so that the column and row specifications for the data are entered into the **Input Range** box as in the preceding box. Then click **OK**. The output from the ANOVA will appear on a new worksheet, as seen next.

	A	B	C	D	E	F	G	H
1	Anova: Single Factor							
2								
3	SUMMARY							
4	*Groups*	*Count*	*Sum*	*Average*	*Variance*			
5	Column 1	8	32	4	3.428571			
6	Column 2	8	44	5.5	2			
7	Column 3	8	64	8	3.428571			
8								
9								
10	ANOVA							
11	*Source of Variation*	*SS*	*df*	*MS*	*F*	*P-value*	*F crit*	
12	Between Groups	65.33333	2	32.66667	11.06452	0.000523	3.4668	
13	Within Groups	62	21	2.952381				
14								
15	Total	127.3333	23					
16								
17								

You can see from the **ANOVA** table that $F(2, 21) = 11.06$, $p = .000523$. In addition to the full **ANOVA** table, Excel also provides the mean and variance for each condition.

USING SPSS

Now, we'll use the data from Table 9.1 to illustrate the use of SPSS to compute a one-way randomized ANOVA. In the study reported in the table, we had participants use one of three different types of rehearsal (rote, imagery, or story) and then had them perform a recall task. Consequently, we manipulated rehearsal and measured memory for the 10 words the participants studied.

Because there were different participants in each condition, we use a randomized ANOVA.

We begin by entering the data into SPSS. The first column is labeled **Rehearsal type** and indicates which type of rehearsal the participants used (1 for rote, 2 for imagery, and 3 for story). The recall data for each of the three conditions appear in the second column, labeled **Recall**.

File	Edit	View	Data	Transform	Analyze	Graphs

27 : Rehearsaltype

	Rehearsaltype	Recall	var
1	1.00	2.00	
2	1.00	4.00	
3	1.00	3.00	
4	1.00	5.00	
5	1.00	2.00	
6	1.00	7.00	
7	1.00	6.00	
8	1.00	3.00	
9	2.00	4.00	
10	2.00	5.00	
11	2.00	7.00	
12	2.00	6.00	
13	2.00	5.00	
14	2.00	4.00	
15	2.00	8.00	
16	2.00	5.00	
17	3.00	6.00	
18	3.00	5.00	
19	3.00	9.00	
20	3.00	10.00	
21	3.00	8.00	
22	3.00	7.00	
23	3.00	10.00	
24	3.00	9.00	
25			

Next, click on **Analyze**, followed by **Compare Means**, and then **One–Way ANOVA** as illustrated in the following.

You should receive the following dialog box.

Enter Rehearsaltype into the **Factor** box by highlighting it and using the appropriate arrow. Do the same for Recall by entering it into the **Dependent List** box. After doing so, the dialog box should appear as follows.

Next, click on the **Options** button and select **Descriptive** and **Continue**. Then click on the **Post Hoc** button and select **Tukey** and then **Continue**. Then click on **OK**. The output from the ANOVA will appear in a new output window as seen next.

Oneway

Descriptives

Recall

	N	Mean	Std. Deviation	Std. Error	95% Confidence Interval for Mean		Minimum	Maximum
					Lower Bound	Upper Bound		
1	8	4.0000	1.85164	.65465	2.4520	5.5480	2.00	7.00
2	8	5.5000	1.41421	.50000	4.3177	6.6823	4.00	8.00
3	8	8.0000	1.85164	.65465	6.4520	9.5480	5.00	10.00
Total	24	5.8333	2.35292	.48029	4.8398	6.8269	2.00	10.00

ANOVA

Recall

	Sum of Squares	df	Mean Square	F	Sig.
Between Groups	65.333	2	32.667	11.065	.001
Within Groups	62.000	21	2.952		
Total	127.333	23			

Post Hoc Tests

Multiple Comparisons

Recall
Tukey HSD

(I) Rehearsaltype	(J) Rehearsaltype	Mean Difference (I-J)	Std. Error	Sig.	95% Confidence Interval	
					Lower Bound	Upper Bound
1.00	2.00	-1.50000	.85912	.212	-3.6655	.6655
	3.00	-4.00000*	.85912	.000	-6.1655	-1.8345
2.00	1.00	1.50000	.85912	.212	-.6655	3.6655
	3.00	-2.50000*	.85912	.022	-4.6655	-.3345
3.00	1.00	4.00000*	.85912	.000	1.8345	6.1655
	2.00	2.50000*	.85912	.022	.3345	4.6655

*. The mean difference is significant at the 0.05 level.

You can see that the descriptive statistics for each condition are provided, followed by the **ANOVA** summary table in which $F(2, 21) = 11.065$, $p = .0005$ (we divide the two-tailed level provided by SPSS in half to obtain the one-tailed level of significance). In addition to the full **ANOVA** summary table, SPSS also calculates Tukey's HSD and provides all pair wise comparisons between the three conditions along with whether or not the comparison was significant.

USING THE TI-84

Let's use the data from Table 9.1 to conduct the analysis using the TI-84 calculator:

1. With the calculator on, press the **STAT** key.
2. **EDIT** will be highlighted. Press the **ENTER** key.
3. Under **L1** enter the data from Table 9.1 for the rote group.
4. Under **L2** enter the data from Table 9.1 for the imagery group.
5. Under **L3** enter the data from Table 9.1 for the story group.
6. Press the **STAT** key once again and highlight **TESTS**.
7. Scroll down to **ANOVA**. Press the **ENTER** key.

8. Next to **ANOVA** enter **(L1,L2,L3)** using the 2nd function key with the appropriate number keys. Make sure that you use commas. The finished line should read "ANOVA(L1,L2,L3)".

9. Press **ENTER**.

The F score of 11.065 should be displayed followed by the significance level of .0005.

MODULE EXERCISES

(Answers appear in Appendix.)

1. A researcher conducts a study on the effects of amount of sleep on creativity. The creativity scores for four levels of sleep (2 hours, 4 hours, 6 hours, and 8 hours) for $n = 5$ participants (in each group) are presented below.

Creativity Scores for Different Sleep Levels
© Cengage Learning 2013

2	4	6	8
3	4	10	10
5	7	11	13
6	8	13	10
4	3	9	9
2	2	10	10

Conduct a one-way randomized ANOVA using Excel, SPSS, and the TI-84 calculator.

2. In a study on the effects of stress on illness, a researcher tallied the number of colds people contracted during a six-month period as a function of the amount of stress they reported during the same time period. There were three stress levels: minimal, moderate, and high. The numbers of colds for the participants in each condition are presented below.

Numbers of Colds for Different Stress Levels
© Cengage Learning 2013

Minimal Stress	Moderate Stress	High Stress
3	2	3
2	3	5
1	2	5
3	4	4
2	3	6

Conduct a one-way randomized ANOVA using Excel, SPSS, and the TI-84 calculator.

Module 10

✳

One-way Repeated Measures ANOVA

Like between-subjects designs, correlated-groups designs may also use more than two levels of an independent variable. You should remember from Module 8 that there are two types of correlated-groups designs: a within-subjects design and a matched-subjects design. The same statistical analyses are used for both designs. We will use a within-subjects design to illustrate the statistical analysis appropriate for a correlated-groups design with more than two levels of an independent variable.

Imagine that we want to conduct the same study described in Module 9 on the effects of rehearsal type on memory, but now using a within-subjects rather than a between-subjects design. In this study, the same three conditions will be used: rote rehearsal, rehearsal with imagery, and rehearsal with a story. The only difference is that the same eight participants serve in every condition. Obviously, we cannot use the same list of words across conditions because there could be a large practice effect. We therefore have to use three lists of words that are equivalent in difficulty and that are counterbalanced across conditions. In other words, not all participants in each condition will receive the same list of words. Let's assume that we have taken the design problems into account and that the data in Table 10.1 represent the performance of the participants in this study. The number of words recalled in each condition is out of 10 words.

T A B L E 10.1   Number of Words Recalled in a Within-subjects Study of the Effects of Rehearsal Type on Memory
© Cengage Learning 2013

Rote Rehearsal	Imagery	Story
2	4	5
3	2	3
3	5	6
3	7	6
2	5	8
5	4	7
6	8	10
4	5	9

You can see that the data are similar to those from the between-subjects design described in Module 9. Because of the similarity in the data, we will be able to see how the statistics used with a within-subjects design are more powerful than those used with a between-subjects design. Because we have interval-ratio data, we will once again use an ANOVA to analyze these data. The only difference will be that the ANOVA used in this case is a *one-way repeated measures ANOVA*. The phrase *repeated measures* refers to the fact that measures were taken repeatedly on the same individuals, that is, the same participants served in all conditions. The difference between this ANOVA and the one-way randomized ANOVA is that the conditions are correlated (related); therefore, the ANOVA procedure must be modified to take this relationship into account.

USING EXCEL

To utilize Excel, we begin by entering the data into Excel, with the data from each condition appearing in a different column and the data for each participant appearing in the same row. Remember that the data for each participant have to appear in a single row. To emphasize this requirement, I've used the first column to indicate each subject in the study, followed by that person's data in the corresponding row, as seen in the following image.

	A	B	C	D
1	Subject	Rote	Imagery	Story
2	1	2	4	5
3	2	3	2	3
4	3	3	5	6
5	4	3	7	6
6	5	2	5	8
7	6	5	4	7
8	7	6	8	10
9	8	4	5	9

Next, with the **Data** ribbon highlighted, click on the **Data Analysis** tab at the top right corner. You will receive the following dialog box.

Select **Anova: Two-Factor Without Replication**, as in the preceding box, and click **OK**. The following dialog box will appear.

```
Anova: Two-Factor Without Replication                    [?][X]

┌─Input───────────────────────────────┐   ┌──────────┐
│ Input Range:        [$B$2:$D$9  ][▦] │   │    OK    │
│                                      │   └──────────┘
│  ☐ Labels                            │   ┌──────────┐
│ Alpha:  [0.05  ]                     │   │  Cancel  │
│                                      │   └──────────┘
└──────────────────────────────────────┘   ┌──────────┐
┌─Output options──────────────────────┐   │   Help   │
│ ○ Output Range:     [          ][▦]  │   └──────────┘
│ ◉ New Worksheet Ply: [          ]    │
│ ○ New Workbook                       │
└──────────────────────────────────────┘
```

With the cursor in the **Input Range** box, highlight the three columns of data so that the column and row specifications for the data are entered into the **Input Range** box as in the preceding window. Make sure you highlight only the data in Columns B, C, and D. Then click **OK**. The output from the ANOVA will appear on a new worksheet as seen in the following.

| File | Home | Insert | Page Layout | Formulas | Data | Review | View |

M15

Anova: Two-Factor Without Replication

	A	B	C	D	E	F	G	H
1	Anova: Two-Factor Without Replication							
2								
3	SUMMARY	Count	Sum	Average	Variance			
4	Row 1	3	11	3.666667	2.333333			
5	Row 2	3	8	2.666667	0.333333			
6	Row 3	3	14	4.666667	2.333333			
7	Row 4	3	16	5.333333	4.333333			
8	Row 5	3	15	5	9			
9	Row 6	3	16	5.333333	2.333333			
10	Row 7	3	24	8	4			
11	Row 8	3	18	6	7			
12								
13	Column 1	8	28	3.5	2			
14	Column 2	8	40	5	3.428571			
15	Column 3	8	54	6.75	5.071429			
16								
17								
18	ANOVA							
19	Source of Variation	SS	df	MS	F	P-value	F crit	
20	Rows	52.5	7	7.5	5	0.00514	2.764199	
21	Columns	42.33333	2	21.16667	14.11111	0.000441	3.738892	
22	Error	21	14	1.5				
23								
24	Total	115.8333	23					
25								

You can see from the **ANOVA** table provided by Excel that $F(2, 14) = 14.11$, $p = .000441$. In addition to the full **ANOVA** table, Excel also provides the mean and variance for each row (subject) and each condition. Moreover, there is also an F score reported for the **Rows** term (the subjects). We report only the F score for the **Columns** (this represents between-groups variance/within-groups variance); the F score for the **Rows** can be ignored.

USING SPSS

(Please note: In order to conduct this analysis, you need the advanced package in addition to the general SPSS package.)

We'll use the data from Table 10.1 to illustrate the use of SPSS to compute a one-way repeated measures ANOVA. As in the Excel example earlier in this module, we will use the same participants in each rehearsal condition, and we therefore will use a repeated measures ANOVA.

We begin by entering the data into SPSS, with the data from each condition appearing in a different column. Remember that the data for each participant have to appear in a single row. To emphasize this requirement, I've used the first column to indicate each subject in the study, followed by that person's data in the corresponding row.

File	Edit	View	Data	Transform	Analyze	Graphs	Utilities	Add-ons	Window	Help

1 : Subject		1.00						
	Subject	Rote	Imagery	Story	var	var	var	var
1	1.00	2.00	4.00	5.00				
2	2.00	3.00	2.00	3.00				
3	3.00	3.00	5.00	6.00				
4	4.00	3.00	7.00	6.00				
5	5.00	2.00	5.00	8.00				
6	6.00	5.00	4.00	7.00				
7	7.00	6.00	8.00	10.00				
8	8.00	4.00	5.00	9.00				
9								
10								
11								
12								

Next, click on **Analyze**, followed by **General Linear Model**, and then **Repeated Measures**. The following dialog box will be produced.

To utilize this dialog box, click in the box beneath **Within–Subject Factor Name**. This represents the independent variable in the study, so enter the name of the independent variable, in this case, Rehearsaltype. Then, in the box below this, let SPSS know how many levels there are to the independent variable. in this case, three. Next click on the **Add** button, which should be active once you've accomplished the two previous steps. We now have to indicate to SPSS what the three levels of Rehearsaltype are. We do this by clicking on **Define**, which should produce the following dialog box.

We now must enter the three levels of Rehearsaltype into slots in the **Within-Subjects Variables** box. Enter this information by highlighting the type of rehearsal and then utilizing the right pointing arrow to enter it into the box. Once all three levels of Rehearsaltype have been entered, click **Options**. This will produce the following dialog box.

Select **Descriptive statistics** and then tell SPSS what variable to calculate descriptive statistics on—Rehearsaltype—by moving this variable into the **Display Means For** box. Click **Continue** and then **OK**. The output will appear in the output sheet. I included only the output necessary to interpret the ANOVA.

General Linear Model

Within-Subjects Factors

Measure:MEASURE_1

Rehearsaltype	Dependent Variable
1	Rote
2	Imagery
3	Story

Descriptive Statistics

	Mean	Std. Deviation	N
Rote	3.5000	1.41421	8
Imagery	5.0000	1.85164	8
Story	6.7500	2.25198	8

Tests of Within-Subjects Effects

Measure:MEASURE_1

Source		Type III Sum of Squares	df	Mean Square	F	Sig.
Rehearsaltype	Sphericity Assumed	42.333	2	21.167	14.111	.000
	Greenhouse-Geisser	42.333	1.943	21.791	14.111	.001
	Huynh-Feldt	42.333	2.000	21.167	14.111	.000
	Lower-bound	42.333	1.000	42.333	14.111	.007
Error(Rehearsaltype)	Sphericity Assumed	21.000	14	1.500		
	Greenhouse-Geisser	21.000	13.599	1.544		
	Huynh-Feldt	21.000	14.000	1.500		
	Lower-bound	21.000	7.000	3.000		

A legend for the three conditions appears first, followed by descriptive statistics for the three conditions. An ANOVA summary table follows. We are concerned only with the data reported for the rows in which sphericity is assumed; these rows represent the standard repeated measures ANOVA. Thus, for these rows, $F(2, 14) = 14.111$, $p = .000$.

If you look back to Module 9—the resulting one-way randomized ANOVA, with very similar data to the repeated measures ANOVA—you can see how much more powerful the repeated measures ANOVA is than the randomized ANOVA. Notice that although the total sums of squares are very similar, the resulting *F*-ratio for the repeated measures ANOVA is much larger (14.11 versus 11.065). If the *F* score is larger, there is a greater probability that it will be statistically significant. Notice also that although the data used to calculate the two ANOVAs are similar, the group means in the repeated measures ANOVA are more similar (closer together) than those from the randomized ANOVA, yet the *F* score from the repeated measures ANOVA is larger. Thus, with somewhat similar data, the resulting *F*-ratio for the repeated measures ANOVA is larger and therefore affords more statistical power.

USING THE TI-84

The TI-84 does not compute a one-way repeated measures ANOVA.

MODULE EXERCISES

(Answers appear in Appendix.)

1. A researcher is interested in the effects of practice on accuracy in a signal detection task. Participants are tested with no practice, after 1 hour of practice, and after 2 hours of practice. Each person participates in all three conditions. The data below indicate how many signals were accurately detected by each participant at each level of practice.

Signal Detection Score Based on Amount of Practice
© Cengage Learning 2013

Participant	No Practice	1 Hour	2 Hours
1	3	4	6
2	4	5	5
3	2	3	4
4	1	3	5
5	3	6	7
6	3	4	6
7	2	3	4

Conduct a one-way repeated-measures ANOVA using Excel and SPSS.

2. A researcher has been hired by a pizzeria to determine which type of crust is the most preferred by customers. The restaurant offers three types of crust: hand-tossed, thick, and thin. The following are the scores for each condition from 10 participants who tasted each type of crust and rated each on a 1 to 5 scale, with 5 being the highest rating.

Signal Detection Score Based on Amount of Practice
© Cengage Learning 2013

Participant	Hand-tossed	Thick	Thin
1	3	3	4
2	4	5	5
3	2	3	4
4	1	2	3
5	3	4	3
6	3	4	4
7	2	3	2
8	1	3	4
9	3	3	5
10	2	2	3

Conduct a one-way repeated-measures ANOVA using Excel and SPSS.

Module 11

✳

Two-way Randomized ANOVA

A two-way ANOVA is similar to a one-way ANOVA in that it analyzes the variance between groups and within groups; however, we use a two-way ANOVA when we have two independent variables. As with the one-way ANOVA, an *F*-ratio is formed by dividing the between-groups variance by the within-groups variance. The difference is that in the two-way ANOVA, between-groups variance may be attributable to Factor A (one of the independent variables in astudy), to Factor B (the second independent variable in astudy), and to the interaction of Factors A and B. With two independent variables, there is a possibility of a main effect for each variable, and an *F*-ratio is calculated to represent each of these effects. In addition, there is the possibility of one interaction effect, and an *F*-ratio is also needed to represent this effect. Thus, with a two-way ANOVA, there are three *F*-ratios to calculate and ultimately to interpret.

In a 2 × 2 factorial design, there are three null and alternative hypotheses. The null hypothesis for Factor A states that there is no effect for Factor A, and the alternative hypothesis states that there is an effect of Factor A (the differences observed between the groups are greater than what would be expected based on chance). In other words, the null hypothesis states that the population means represented by the sample means are from the same population, whereas the alternative hypothesis states that the population means represented by the sample means are not from the same population. A second null hypothesis states that there is no effect for Factor B, whereas the alternative hypothesis states that there is an effect of Factor B. The third null hypothesis states that there is no interaction of Factors A and B, whereas the alternative hypothesis states that there is an interaction effect.

Let's use a memory study to illustrate the calculation of a two-way randomized ANOVA. In this study, Factor A is word type (concrete versus abstract), Factor B is rehearsal type (rote versus imagery), and the dependent variable is the number of words recalled. Thus, there are two independent variables (word type and rehearsal type), each with two levels. Table 11.1 presents the number of words recalled by the 32 participants in the memory study with 8 participants in each condition.

TABLE 11.1   Number of Words Recalled as a Function of Word Type and Rehearsal Type
© Cengage Learning 2013

	Word Type	
Rehearsal Type	**Abstract**	**Concrete**
Rote	5	4
	4	5
	5	3
	6	6
	4	2
	5	2
	6	6
	5	4
Imagery	6	10
	5	12
	6	11
	7	9
	6	8
	6	10
	7	10
	5	10

USING EXCEL

We'll use the data from Table 11.1 to illustrate the use of Excel to compute a two-way randomized ANOVA. This study is similar to the previous two studies for the one-way ANOVAs, except that we are introducing a second independent variable (word type). Accordingly, there are two types of rehearsal (rote versus imagery) and two word types that participants might study (abstract versus concrete). Because we used different participants in each condition, we use a randomized ANOVA, and because there are two independent variables, the ANOVA is two–way.

We begin by entering the data into Excel with the column headings for word type and the row headings for rehearsal type included. See the example that follows.

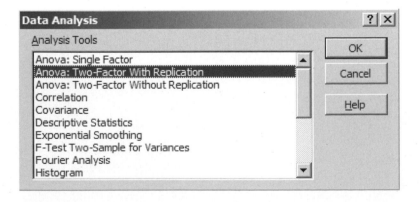

	Abstract Words	Concrete Words
Rote Rehearsal	5	4
	4	5
	5	3
	6	6
	4	2
	5	2
	6	6
	5	4
Imagery Rehearsal	6	10
	5	12
	6	11
	7	9
	6	8
	6	10
	7	10
	5	10

Next, with the **Data** ribbon highlighted, click on the **Data Analysis** tab at the top right corner. You will receive the following dialog box.

Data Analysis

Analysis Tools

- Anova: Single Factor
- Anova: Two-Factor With Replication
- Anova: Two-Factor Without Replication
- Correlation
- Covariance
- Descriptive Statistics
- Exponential Smoothing
- F-Test Two-Sample for Variances
- Fourier Analysis
- Histogram

OK   Cancel   Help

Select **Anova: Two-Factor With Replication** and click **OK**. The following dialog box will appear.

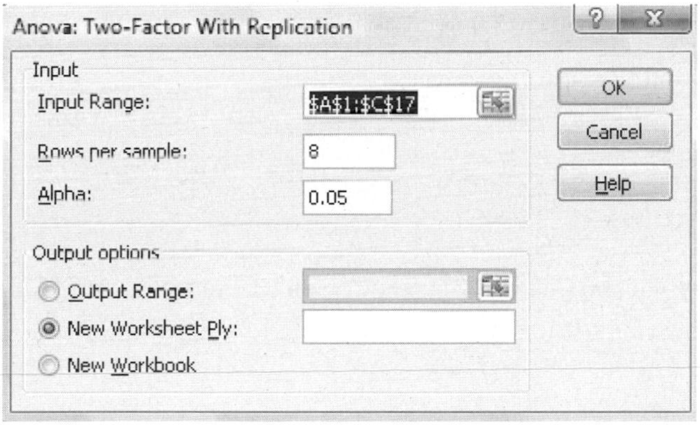

With the cursor in the **Input Range** box, highlight the three columns of labels and data so that the column and row specifications for the data are entered into the **Input Range** box as in the preceding window. Next enter the number of **Rows per sample**, which in this case is 8. Then click **OK**. The output from the ANOVA will appear on a new worksheet as seen in the following image.

| File | Home | Insert | Page Layout | Formulas | Data | Review | View |

| H35 | f_x |

	A	B	C	D	E	F	G	H
1	Anova: Two-Factor With Replication							
2								
3	SUMMARY	Abstract Words	Concrete Words	Total				
4	*Rote Rehearsal*							
5	Count	8	8	16				
6	Sum	40	32	72				
7	Average	5	4	4.5				
8	Variance	0.571428571	2.571428571	1.733333				
9								
10	*Imagery Rehearsal*							
11	Count	8	8	16				
12	Sum	48	80	128				
13	Average	6	10	8				
14	Variance	0.571428571	1.428571429	5.2				
15								
16	*Total*							
17	Count	16	16					
18	Sum	88	112					
19	Average	5.5	7					
20	Variance	0.8	11.46666667					
21								
22								
23	ANOVA							
24	*Source of Variation*	*SS*	*df*	*MS*	*F*	*P-value*	*F crit*	
25	Sample	98	1	98	76.22222	1.76E-09	4.195972	
26	Columns	18	1	18	14	0.000836	4.195972	
27	Interaction	50	1	50	38.88889	9.72E-07	4.195972	
28	Within	36	28	1.285714				
29								
30	Total	202	31					
31								

In the preceding **ANOVA** table, what is called "Sample" is the independent variable of rehearsal type, what is called "Columns" is the independent variable of word type, and the interaction is labeled as such. We see that the three F scores are all significant. In addition, the means and variances for each condition, and summarized across conditions, are reported. Thus, there is an effect of rehearsal type, $F(1, 28) = 76.22$, $p = .000000002$, with participants recalling more words in the imagery rehearsal condition. There is also an effect of word type, $F(1, 28) = 14$, $p = .0008$, with participants recalling significantly more concrete words than abstract words. Lastly, there is an interaction effect, $F(1, 28) - 38.89$, $p = .00000097$, showing that participants recalled abstract words somewhat similarly across the two rehearsal type conditions but that they recalled concrete words much better in the imagery rehearsal condition in comparison to the rote rehearsal condition.

USING SPSS

We'll use the data from Table 11.1 to illustrate the use of SPSS to compute a two-way randomized ANOVA. We begin by entering the data into SPSS as indicated in the Data Sheet below. The first two columns indicate the condition in which the participants served with 1,1 indicating the abstract word, rote rehearsal condition; 1,2 the abstract word, imagery rehearsal condition; 2,1 the concrete word, rote rehearsal condition; and 2,2 the concrete word, imagery rehearsal condition. The recall data for each of the participants in these four conditions are entered into the third column.

File   Edit   View   Data   Transform   Analyze   Graphs   Utilities   Add-ons   Window   Help

	Wordtype	Rehearsaltype	Wordsrecalled	var	var	var	var	var	var	var
1	1.00	1.00	5.00							
2	1.00	1.00	4.00							
3	1.00	1.00	5.00							
4	1.00	1.00	6.00							
5	1.00	1.00	4.00							
6	1.00	1.00	5.00							
7	1.00	1.00	6.00							
8	1.00	1.00	5.00							
9	1.00	2.00	6.00							
10	1.00	2.00	5.00							
11	1.00	2.00	6.00							
12	1.00	2.00	7.00							
13	1.00	2.00	6.00							
14	1.00	2.00	6.00							
15	1.00	2.00	7.00							
16	1.00	2.00	5.00							
17	2.00	1.00	4.00							
18	2.00	1.00	5.00							
19	2.00	1.00	3.00							
20	2.00	1.00	6.00							
21	2.00	1.00	2.00							
22	2.00	1.00	2.00							
23	2.00	1.00	6.00							
24	2.00	1.00	4.00							
25	2.00	2.00	10.00							
26	2.00	2.00	12.00							
27	2.00	2.00	11.00							
28	2.00	2.00	9.00							
29	2.00	2.00	8.00							
30	2.00	2.00	10.00							
31	2.00	2.00	10.00							
32	2.00	2.00	10.00							
33										
34										
35										

Data View   Variable View

Next, click on **Analyze,** followed by **General Linear Model**, and **Univariate** as indicated below.

File	Edit	View	Data	Transform	Analyze	Graphs	Utilities	Add-ons	Window	Help

Reports ▶
Descriptive Statistics ▶
Compare Means ▶
General Linear Model ▶   Univariate...
Correlate ▶
Regression ▶
Classify ▶
Dimension Reduction ▶
Scale ▶
Nonparametric Tests ▶
Forecasting ▶
Multiple Response ▶
Quality Control ▶
ROC Curve...

	Wordtype	Rehears			var
1	1.00				
2	1.00				
3	1.00				
4	1.00				
5	1.00				
6	1.00				
7	1.00				
8	1.00				
9	1.00				
10	1.00				
11	1.00	2.00	6.00		
12	1.00	2.00	7.00		
13	1.00	2.00	6.00		
14	1.00	2.00	6.00		
15	1.00	2.00	7.00		
16	1.00	2.00	5.00		

The following dialog box will be produced.

Enter the dependent variable (Wordsrecalled) into the **Dependent Variable** box and the two independent variables (Wordtype and Rehearsaltype) into the **Fixed Factors** box by highlighting each variable and utilizing the appropriate arrow keys. Next select **Options**, which will produce the following dialog box.

Select **Descriptive statistics** and then tell SPSS that you want descriptive statistics on all factors by moving OVERALL into the **Display Means for** box. Click **Continue** and then **OK**. The output will be displayed in an output window as seen in the following image.

Univariate Analysis of Variance

Between-Subjects Factors

		N
Wordtype	1.00	16
	2.00	16
Rehearsaltype	1.00	16
	2.00	16

Descriptive Statistics

Dependent Variable:Wordsrecalled

Wordtype	Rehearsaltype	Mean	Std. Deviation	N
1.00	1.00	4.0000	1.60357	8
	2.00	10.0000	1.19523	8
	Total	7.0000	3.38625	16
2.00	1.00	5.0000	.75593	8
	2.00	6.0000	.75593	8
	Total	5.5000	.89443	16
Total	1.00	4.5000	1.31656	16
	2.00	8.0000	2.28035	16
	Total	6.2500	2.55267	32

Tests of Between-Subjects Effects

Dependent Variable:Wordsrecalled

Source	Type III Sum of Squares	df	Mean Square	F	Sig.
Corrected Model	166.000[a]	3	55.333	43.037	.000
Intercept	1250.000	1	1250.000	972.222	.000
Wordtype	18.000	1	18.000	14.000	.001
Rehearsaltype	98.000	1	98.000	76.222	.000
Wordtype * Rehearsaltype	50.000	1	50.000	38.889	.000
Error	36.000	28	1.286		
Total	1452.000	32			
Corrected Total	202.000	31			

a. R Squared = .822 (Adjusted R Squared = .803)

The output begins with a legend for the variables and then is followed by descriptive statistics. The ANOVA summary table follows. The rows that correspond to the standard two-way ANOVA are those labeled **Wordtype** through **Total**. We can see that the three F-ratios reported in these rows correspond to those from the Excel example and that they can be interpreted in the same manner as they were earlier in the module. There is an effect of rehearsal type, $F(1, 28) = 76.22$, $p - .000$, with participants recalling more words in the imagery rehearsal condition. There is also an effect of word type, $F(1, 28) = 14$, $p = .001$ with participants recalling significantly more concrete words than abstract words. Finally, there is an interaction effect, $F(1, 28) = 38.89$, $p = .000$, showing that participants recalled abstract words somewhat similarly across the two rehearsal type conditions but that they recalled concrete words much better in the imagery rehearsal condition in comparison to the rote rehearsal condition.

MODULE EXERCISES

(Answers to odd-numbered questions appear in Appendix.)

1. In a study, a researcher measures the preference of males and females for two brands of frozen pizza (one low fat and one regular). The table below shows the preference scores on a 10-point scale for each of the 24 participants in the study.

	Females	Males
Brand 1	3	9
(Low Fat)	4	7
	2	6
	2	8
	5	9
	3	7
Brand 2	8	4
(Regular)	9	2
	7	5
	10	6
	9	2
	10	5

© Cengage Learning 2013

Conduct a two-way randomized ANOVA using Excel and SPSS.

2. A researcher is attempting to determine the effects of practice and gender on a timed task. Participants in an experiment were given a computerized search task. They searched a computer screen of various characters and attempted to find a particular character on each trial. When they found the designated character, they pressed a button stopping a timer. Their reaction time (in seconds) on each trial was recorded. Participants practiced for either 2, 4, or 6 hours and were either female or male. The reaction time data for the 30 participants appear below.

	Females	Males
2 Hours	12	11
	13	12
	12	13
	11	12
	11	11
4 Hours	10	8
	10	8
	10	10
	8	10
	7	9
6 Hours	7	5
	5	6
	7	8
	6	6
	7	8

Conduct a two-way randomized ANOVA using Excel and SPSS.

CHAPTER 6

✳

Correlational Procedures

In this chapter, we discuss correlational procedures. Correlational designs allow us to describe the relationship between two measured variables. A correlation coefficient aids us by assigning a numerical value to the observed relationship. We begin with the procedure for calculating one of the more commonly used correlation coefficients, the Pearson product-moment correlation coefficient. We then introduce simple regression analysis.

Module 12

✳

Pearson Product-Moment Correlation Coefficient

The most commonly used correlation coefficient is the *Pearson product-moment correlation coefficient*, usually referred to as Pearson's *r* (*r* is the statistical notation we use to report correlation coefficients). Pearson's *r* is used for data measured on an interval or ratio scale of measurement. Refer to Table 12.1, which presents the height and weight data for 20 individuals. Because height and weight are both measured on a ratio scale, Pearson's *r* would be applicable to these data.

TABLE 12.1   Height and Weight Data for 20 Individuals
© Cengage Learning 2013

Weight	Height
100	60
120	61
105	63
115	63
119	65
134	65
129	66
143	67
151	65
163	67
160	68
176	69
165	70
181	72
192	76
208	75
200	77
152	68
134	66
138	65

The development of this correlation coefficient is typically credited to Karl Pearson (hence the name), who published his formula for calculating r in 1895. Actually, Francis Edgeworth published a similar formula for calculating r in 1892. Not realizing the significance of his work, however, Edgeworth embedded the formula in a statistical paper that was very difficult to follow, and it was not noted until years later. Thus, although Edgeworth had published the formula three years earlier, Pearson received the recognition.

USING EXCEL

To illustrate how Excel can be used to calculate a correlation coefficient, let's use the data from Table 12.1 from which we will calculate Pearson's product-moment correlation coefficient. In order to make this calculation, we begin by entering the data from Table 12.1 into Excel, as seen in the following figure in which the weight data were entered into Column A and the height data into Column B.

Next, with the **Data** ribbon active, as in the preceding window, click on **Data Analysis** at the upper right corner. The following dialog box will appear.

Highlight **Correlation** and then click **OK**. The subsequent dialog box will appear.

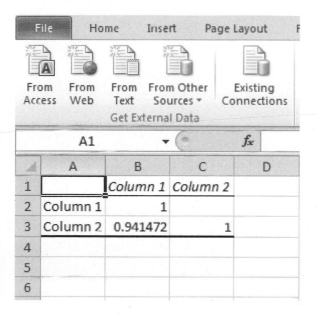

With the cursor in the **Input Range** box, highlight the data in Columns A and B and click **OK**. The output worksheet generated from this operation is very small and simply reports the correlation coefficient of +.94, as seen in the following.

USING SPSS

To illustrate how SPSS can be used to calculate a correlation coefficient, let's use the data from Table 12.1 from which we will calculate Pearson's product–moment correlation coefficient, just as we did above. In order to make this calculation, we begin by entering the data from Table 12.1 into SPSS, as seen in the following image in which the weight data were entered into Column A and the height data into Column B.

File	Edit	View	Data	Transform	Analyze	Graphs	Utilities	Ad

	Weight	Height	var	var
1	100.00	60.00		
2	120.00	61.00		
3	105.00	63.00		
4	115.00	63.00		
5	119.00	65.00		
6	134.00	65.00		
7	129.00	66.00		
8	143.00	67.00		
9	151.00	65.00		
10	163.00	67.00		
11	160.00	68.00		
12	176.00	69.00		
13	165.00	70.00		
14	181.00	72.00		
15	192.00	76.00		
16	208.00	75.00		
17	200.00	77.00		
18	152.00	68.00		
19	134.00	66.00		
20	138.00	65.00		
21				

Next, click on **Analyze,** followed by **Correlate**, and then **Bivariate**. The dialog box that follows will be produced.

Move the two variables you want correlated (Weight and Height) into the **Variables** box. In addition, click **One-tailed** because this was a one-tailed test and then click on **Options** and select **Means and standard deviations**, thus letting SPSS know that you want descriptive statistics on the two variables. The dialog box should now appear as follows.

Click **OK** to receive the following output.

Correlations

Descriptive Statistics

	Mean	Std. Deviation	N
Weight	149.2500	31.21213	20
Height	67.4000	4.68368	20

Correlations

		Weight	Height
Weight	Pearson Correlation	1	.941**
	Sig. (1-tailed)		.000
	N	20	20
Height	Pearson Correlation	.941**	1
	Sig. (1-tailed)	.000	
	N	20	20

**. Correlation is significant at the 0.01 level (1-tailed).

The correlation coefficient of .941 is provided along with the one-tailed significance level and the mean and standard deviation for each of the variables.

USING THE TI-84

Let's use the data from Table 12.1 to conduct the analyses using the TI-84 calculator:

1. With the calculator on, press the **STAT** key.
2. **EDIT** will be highlighted. Press the **ENTER** key.
3. Under **L1** enter the weight data from Table 12.1.
4. Under **L2** enter the height data from Table 12.1.
5. Press the 2nd key and 0 [catalog] and scroll down to **DiagnosticOn** and press **ENTER**. Press **ENTER** once again. (The message **DONE** should appear on the screen.)
6. Press the **STAT** key and highlight **CALC**. Scroll down to **8:LinReg(a+bx)** and press **ENTER**.
7. Next to **LinReg(a+bx)**, type **L1** (by pressing the 2nd key followed by the **1** key) followed by a comma and then **L2** (by pressing the 2nd key followed by the **2** key). On the screen should appear **LinReg(a+bx) L1,L2**
8. Press **ENTER**.

The values of a (46.31), b (0.141), r^2 (.89), and r (.94) should appear on the screen. You can see that r (the correlation coefficient) is the same as that calculated by Excel and SPSS.

MODULE EXERCISES

(Answers appear in Appendix.)

1. In a study of caffeine and stress, college students indicate how many cups of coffee they drink per day and their stress level on a scale from 1 to 10. The data appear below.

Number of Cups of Coffee	Stress Level
3	5
2	3
4	3
6	9
5	4
1	2
7	10
3	5

Calculate a Pearson's *r* using Excel, SPSS, and the TI-84.

2. Given the data below, determine the correlation between IQ scores and psychology exam scores, between IQ scores and statistics exam scores, and between psychology exam scores and statistics exam scores using Excel, SPSS, and the TI-84.

Student	IQ Score	Psychology Exam Score	Statistics Exam Score
1	140	48	47
2	98	35	32
3	105	36	38
4	120	43	40
5	119	30	40
6	114	45	43
7	102	37	33
8	112	44	47
9	111	38	46
10	116	46	44

Module 13

✳

Regression Analysis

The correlational procedure allows you to predict from one variable to another; the degree of accuracy with which you can predict depends on the strength of the correlation. A tool that enables us to predict an individual's score on one variable based on knowing one or more other variables is known as *regression analysis*. For example, imagine that you are an admissions counselor at a university and you want to predict how well a prospective student might do at your school based on both SAT scores and high school GPA. Or imagine that you work in a human resources office and you want to predict how well future employees might perform based on test scores and performance measures. Regression analysis allows you to make such predictions by developing a regression equation.

To illustrate regression analysis, let's use the height and weight data presented in Table 13.1. When we used these data to calculate Pearson's *r* in Module 12, we determined that the correlation coefficient was +.94.

TABLE 13.1   **Height and Weight Data for 20 Individuals**
© Cengage Learning 2013

Weight	Height
100	60
120	61
105	63
115	63
119	65
134	65
129	66
143	67
151	65
163	67
160	68
176	69
165	70
181	72
192	76
208	75
200	77
152	68
134	66
138	65

We can see in the scattergram in Figure 13.1 that there is a linear relationship between the variables, meaning that a straight line can be drawn through the data to represent the relationship between the variables. This *regression line* is shown in the figure, and it represents the relationship between height and weight for this group of individuals.

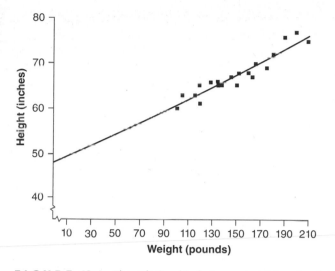

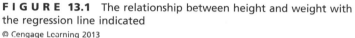

FIGURE 13.1 The relationship between height and weight with the regression line indicated

© Cengage Learning 2013

Regression analysis involves determining the equation for the best-fitting line for a data set. This equation is based on the equation for representing a line that you may remember from algebra class: $y = mx + b$, where m is the slope of the line and b the y-intercept (the place at which the line crosses the y-axis). For a linear regression analysis, the formula is essentially the same, although the symbols differ:

$$Y' = bX + a,$$

where Y' is the predicted value on the Y variable, b is the slope of the line, X represents an individual's score on the X variable, and a is the y-intercept.

Using this formula, then, we can predict an individual's approximate score on variable Y based on that person's score on variable X. With the height and weight data, for example, we could predict an individual's approximate height based on knowing the person's weight. You can picture what we are talking about by looking at Figure 13.1 above. Given the regression line in the Figure, if we know an individual's weight (read from the x-axis), we can then predict the person's height (by finding the corresponding value on the y-axis).

USING EXCEL

To illustrate how Excel can be used to calculate a regression analysis, let's use the data from Table 13.1 from which we will calculate a regression line. In order to make this calculation, we begin by entering the data from Table 13.1 into Excel. The following figure illustrates this operation, with the weight data being entered into Column A and the height data into Column B.

	A	B	C	D	E	F	G	H	I	J	K	L	M	N	O	P	Q
1	100	60															
2	120	61															
3	105	63															
4	115	63															
5	119	65															
6	134	65															
7	129	66															
8	143	67															
9	151	65															
10	163	67															
11	160	68															
12	176	69															
13	165	70															
14	181	72															
15	192	76															
16	208	75															
17	200	77															
18	152	68															
19	134	66															
20	138	65															
21																	

Next, with the **Data** ribbon active, as in the preceding window, click on **Data Analysis** at the upper right corner. The following drop-down box will appear.

Highlight **Regression** and then click **OK**. The dialog box that follows will appear.

With the cursor in the **Input Y Range** box, highlight the height data in Column B so that it appears in the **Input Y Range** box. Do the same with the **Input X Range** box and the data from Column A (we place the height data in the Y box because this is what we are predicting—height—based on knowing one's weight). Then click **OK**. The following output will be produced.

| File | Home | Insert | Page Layout | Formulas | Data | Review | View |

(Excel ribbon: Get External Data — From Access, From Web, From Text, From Other Sources, Existing Connections; Connections — Refresh All, Properties, Edit Links, Connections; Sort & Filter — Sort, Filter, Clear, Reapply, Advanced; Data Tools — Text to Columns, Remove Duplicates, Data Validation, Consoli...)

A1   fx   SUMMARY OUTPUT

	A	B	C	D	E	F	G	H	I	J
1	SUMMARY OUTPUT									
2										
3	*Regression Statistics*									
4	Multiple R	0.941472332								
5	R Square	0.886370152								
6	Adjusted R Squ	0.880057383								
7	Standard Error	1.622085773								
8	Observations	20								
9										
10	ANOVA									
11		*df*	*SS*	*MS*	*F*	*Significance F*				
12	Regression	1	369.4390794	369.4391	140.4091	6.1824E-10				
13	Residual	18	47.3609206	2.631162						
14	Total	19	416.8							
15										
16		*Coefficients*	*Standard Error*	*t Stat*	*P-value*	*Lower 95%*	*Upper 95%*	*Lower 95.0%*	*Upper 95.0%*	
17	Intercept	46.31442348	1.815048149	25.50286	1.4E-15	42.4990479	50.1297991	42.4990479	50.12979906	
18	X Variable 1	0.141276895	0.01192267	11.84943	6.18E-10	0.11622829	0.1663255	0.11622829	0.166325496	
19										

We are primarily interested in the data necessary to create the regression line—the Y-intercept and the slope. These data can be found on lines 17 and 18 of the output worksheet in the first column labeled **Coefficients**. We see that the Y-intercept is 46.31 and the slope is .141. Thus, the regression equation would be $Y' = .141(X) + 46.31$.

USING SPSS

To illustrate how SPSS can be used to calculate a regression analysis, let's again use the data from Table 13.1 from which we will calculate a regression line, just as we did with Excel. In order to make this calculation, we begin by entering the data from Table 13.1 into SPSS. The following figure illustrates this operation, with the data being entered just as they were when we used SPSS to calculate a correlation coefficient in Module 12.

File Edit View Data Transform Analyze Graphs Utilities Ad

	Weight	Height	var	var
1	100.00	60.00		
2	120.00	61.00		
3	105.00	63.00		
4	115.00	63.00		
5	119.00	65.00		
6	134.00	65.00		
7	129.00	66.00		
8	143.00	67.00		
9	151.00	65.00		
10	163.00	67.00		
11	160.00	68.00		
12	176.00	69.00		
13	165.00	70.00		
14	181.00	72.00		
15	192.00	76.00		
16	208.00	75.00		
17	200.00	77.00		
18	152.00	68.00		
19	134.00	66.00		
20	138.00	65.00		
21				

Next, click on **Analyze**, followed by **Regression**, and then **Linear** as in the following dialog box.

File	Ecit	View	Data	Transform	Analyze	Graphs	Utilites	Add-ons	Window	Help

Reports ▶
Descriptive Statistics ▶
Compare Means ▶
General Linear Model ▶
Correlate ▶
Regression ▶
Classify ▶
Dimension Reduction ▶
Scale ▶
Nonparametric Tests ▶
Forecasting ▶
Multiple Response ▶
Quality Control ▶
ROC Curve...

Linear...
Curve Estimation...
Partial Least Squares...
Ordinal...

	Participant	Weig		var	var	var
1	1.00	10				
2	2.00	12				
3	3.00	10				
4	4.00	11				
5	5.00	11				
6	6.00	13				
7	7.00	12				
8	8.00	14				
9	9.00	15				
10	10.00	16				
11	11.00	160.00	68.00			
12	12.00	176.00	69.00			
13	13.00	165.00	70.00			
14	14.00	181.00	72.00			
15	15.00	192.00	76.00			
16	16.00	208.00	75.00			
17	17.00	200.00	77.00			
18	18.00	152.00	68.00			
19	19.00	134.00	66.00			
20	20.00	138.00	65.00			

The dialog box that follows will be produced.

For this regression analysis, we are attempting to predict height based on knowing an individual's weight. Thus, we are using height as the dependent measure in our model and weight as the independent measure. Enter Height into the **Dependent** box and Weight into the **Independent** box by using the appropriate arrows. Then click **OK**. The output will be generated in the output window.

Regression

Variables Entered/Removed[b]

Model	Variables Entered	Variables Removed	Method
1	Weight[a]	.	Enter

a. All requested variables entered.
b. Dependent Variable: Height

Model Summary

Model	R	R Square	Adjusted R Square	Std. Error of the Estimate
1	.941[a]	.886	.880	1.62209

a. Predictors: (Constant), Weight

Coefficients[a]

Model		Unstandardized Coefficients		Standardized Coefficients	t	Sig.
		B	Std. Error	Beta		
1	(Constant)	46.314	1.816		25.503	.000
	Weight	.141	.012	.941	11.849	.000

a. Dependent Variable: Height

We are most interested in the data necessary to create the regression line—the Y-intercept and the slope. These data can be found in the box labeled **Coefficients**. We see that the Y-intercept **(Constant)** is 46.314 and the slope is 0.141. Therefore, the regression equation would be $Y' = 0.141(X) + 46.31$.

USING THE TI-84

Let's use the data from Table 13.1 to conduct the regression analysis using the TI–84 calculator:

1. With the calculator on, press the **STAT** key.
2. **EDIT** will be highlighted. Press the **ENTER** key.
3. Under **L1** enter the weight data from Table 13.1.
4. Under **L2** enter the height data from Table 13.1.
5. Press the 2nd key and 0 [catalog] and scroll down to **DiagnosticOn** and press **ENTER**. Press **ENTER** once again. (The message **DONE** should appear on the screen.)

6. Press the **STAT** key and highlight **CALC**. Scroll down to **8:LinReg(a+bx)** and press **ENTER**.

7. Next to **LinReg(a+bx)**, type **L1** (by pressing the 2nd key followed by the **1** key) followed by a comma and **L2** (by pressing the 2nd key followed by the **2** key). It should appear as follows on the screen. **LinReg(a+bx) L1,L2**

8. Press **ENTER**.

The values of a (46.31), b (0.141), r^2 (.89), and r (.94) should appear on the screen.

MODULE EXERCISES

(Answers to odd-numbered questions appear in Appendix.)

1. In a study of caffeine and stress, college students indicate how many cups of coffee they drink per day and their stress level on a scale from 1 to 10. The data appear below.

Number of Cups of Coffee	Stress Level
3	5
2	3
4	3
6	9
5	4
1	2
7	10
3	5

Determine the regression equation between these two variables.

2. Given the data below, determine the regression equation between IQ scores and psychology exam scores, between IQ scores and statistics exam scores, and between psychology exam scores and statistics exam scores using Excel, SPSS, and the TI-84.

Student	IQ Score	Psychology Exam Score	Statistics Exam Score
1	140	48	47
2	98	35	32
3	105	36	38
4	120	43	40
5	119	30	40
6	114	45	43
7	102	37	33
8	112	44	47
9	111	38	46
10	116	46	44

CHAPTER 7

✳

Nonparametric Statistics

Statistics used to analyze nominal and ordinal data are referred to as *nonparametric tests*. A nonparametric test is a test that does not involve the use of any population parameters. In other words, the mean and standard deviation are not needed, and the underlying distribution does not have to be normal. In addition, most nonparametric tests are based on fewer assumptions than parametric tests and are usually easier to compute. They are, however, less powerful than parametric tests, meaning that it is more difficult to reject the null hypothesis when it is false. In this chapter, we will look at two nonparametric tests for nominal data: the χ^2 goodness-of-fit test and the χ^2 test of independence. We will also discuss two tests for ordinal data: the Wilcoxon rank-sum test, used with between subjects designs, and the Wilcoxon matched pairs signed-ranks test, used with correlated-groups designs.

Module 14

✳

Chi-square Tests

CHI-SQUARE GOODNESS-OF-FIT TEST

The *chi-square (χ^2) goodness-of-fit test* is used for comparing categorical information against what we would expect based on previous knowledge. As such, it tests what are called *observed frequencies* (the frequency with which participants fall into a category) against *expected frequencies* (the frequency expected in a category if the sample data represent the population). It is a nondirectional test, meaning that the alternative hypothesis is neither one-tailed nor two-tailed. The alternative hypothesis for a χ^2 goodness-of-fit test is that the observed data do not fit the expected frequencies for the population, and the null hypothesis is that they do fit the expected frequencies for the population. There is no conventional way to write these hypotheses in symbols, as we have done with the previous statistical tests. To illustrate the χ^2 goodness-of-fit test, let's look at a situation in which its use would be appropriate.

Suppose that a researcher is interested in determining whether the teenage pregnancy rate at a particular high school is different from the rate statewide. Assume that the rate statewide is 17%. A random sample of 80 female students is selected from the target high school. Seven of the students are either pregnant now or have been pregnant previously. The χ^2 goodness-of-fit test measures the observed frequencies against the expected frequencies. The observed and expected frequencies are presented in Table 14.1.

T A B L E 14.1   **Observed and Expected Frequencies for χ^2 Goodness-of-fit Example**
© Cengage Learning 2013

Frequencies	Pregnant	Not Pregnant
Observed	7	73
Expected	14	66

As can be seen in the table, the observed frequencies represent the number of high school females in the sample of 80 who were pregnant versus not pregnant. The expected frequencies represent what we would expect based on chance, given what is known about the population. In this case, we would expect 17% of the females to be pregnant because this is the rate statewide. If we take 17% of 80 ($.17 \times 80 = 14$), we would expect 14 of the students to be pregnant. By the same token, we would expect 83% of the students ($.83 \times 80 = 66$) to be not pregnant. If the calculated expected frequencies are correct, when summed they should equal the sample size ($14 + 66 = 80$).

Using SPSS

To begin using SPSS to calculate this chi-square test, we must enter the data into the Data Editor. We have 80 individuals in the sample, thus there will be 80 scores entered. A score of 1 indicates that the individual was pregnant (there are 7 scores of 1 entered), and a score of 2 indicates that the individual was not pregnant (there are 73 scores of 2 entered). A portion of the Data Editor with the data entered appears next. I've named the variable **Observedpregnancy** to indicate that we are interested in the pregnancy rate observed in the sample.

	File	Edit	View	Data	Transform	Analyze	Graphs	Utilities	Add-ons	Window	Help

	Observedpregnancy	var	var	var	var
1	1				
2	1				
3	1				
4	1				
5	1				
6	1				
7	1				
8	2				
9	2				
10	2				
11	2				
12	2				
13	2				
14	2				
15	2				
16	2				
17	2				
18	2				
19	2				
20	2				
21	2				
22	2				
23	2				
24	2				
25	2				
26	2				
27	2				
28	2				
29	2				
30	2				
31	2				
32	2				
33	2				
34	2				
35	2				

Data View   Variable View

Next, we click on **Analyze**, then **Nonparametric Tests**, **Legacy Dialogs**, and finally **Chi-square**, as illustrated in the following screen capture.

This operation will produce the following dialog box.

Our **Test Variable List** is the data in the Observedpregnancy variable. Thus, we'll move the Observedpregnancy variable over to the **Test Variable List** using the arrow key. We also have to specify the expected values in the **Expected Values** box. Click on **Values** and then type the first value (14) into the box and click on **Add**. Next, enter the second expected value (66) by clicking on **Add**. The dialog box should now appear as follows.

Click **OK**, and you should receive the output that appears next.

NPar Tests

Chi-Square Test

Frequencies

Observedpregnancy

	Observed N	Expected N	Residual
1	7	14.0	-7.0
2	73	66.0	7.0
Total	80		

Test Statistics

	Observedpreg nancy
Chi-square	4.242[a]
df	1
Asymp. Sig.	.039

a. 0 cells (.0%) have expected frequencies less than 5. The minimum expected cell frequency is 14.0.

We can reject the null hypothesis and conclude that the observed frequency of pregnancy is significantly lower than expected by chance. In other words, the female teens at the target high school have a significantly lower pregnancy rate than would be expected based on the statewide rate. In APA style, this would be reported as χ^2 (1, $N = 80$) = 4.24, $p = .039$.

Using the TI-84

Let's use the data from Table 14.1 to conduct the calculation using the TI-84 calculator:

1. With the calculator on, press the **STAT** key.
2. Highlight **EDIT** and press **ENTER**.

3. Enter the observed scores in **L1** and the expected scores in **L2**.
4. Press the **STAT** key and highlight **TESTS**.
5. Scroll down to **D:** χ^2 **GOF Test** and press **ENTER**.
6. The calculator should show **Observed: L1, Expected: L2**, and **df = 1**.
7. Scroll down and highlight **CALCULATE** and press **ENTER**.

The χ^2 value of a (4.24) should appear on the screen, along with the $df = 1$ and $p = .039$.

CHI-SQUARE TEST OF INDEPENDENCE

The logic of the *chi-square* (χ^2) *test of independence* is the same as for any χ^2 statistic: We are comparing how well an observed breakdown of people over various categories fits some expected breakdown (such as an equal breakdown). In other words, a χ^2 test compares an observed frequency distribution to an expected frequency distribution. If we observe a difference, we determine whether the difference is greater than what would be expected based on chance. The difference between the χ^2 test of independence and the χ^2 goodness-of-fit test is that the goodness-of-fit test compares how well an observed frequency distribution of *one* nominal variable fits some expected pattern of frequencies, whereas the test of independence compares how well an observed frequency distribution of *two* nominal variables fits some expected pattern of frequencies. The null hypothesis and the alternative hypothesis are similar to those used with the *t* tests. The null hypothesis is that there are no observed differences in frequency between the groups we are comparing; the alternative hypothesis is that there are differences in frequency between the groups and that the differences are greater than we would expect based on chance.

As a means of illustrating the χ^2 test of independence, imagine that teenagers in a randomly chosen sample are categorized as having been employed as babysitters or never having been employed in this capacity. The teenagers are then asked whether they have ever taken a first aid course. In this case, we would like to determine whether babysitters are more likely to have taken first aid than those who have never worked as babysitters. Because we are examining the observed frequency distribution of two nominal variables (babysitting and taking a first aid class), the χ^2 test of independence is appropriate. We find that

TABLE 14.2   **Observed and Expected Frequencies for Babysitters and Nonbabysitters Having Taken a First Aid Course**
© Cengage Learning 2013

	Taken First Aid	
Babysitter	Yes	No
Yes	12	6
No	5	13

12 of the 18 babysitters have had a first aid course and 6 of the babysitters have not. In the nonbabysitter group, 5 out of 18 have had a first aid course, and the remaining 13 have not. Table 14.2 is a contingency table showing the observed frequencies.

Using SPSS

We begin by entering the data into SPSS. We enter the data by indicating whether the individual is a babysitter (1) or not a babysitter (2) and whether they have taken a first aid class (1) or not (2). Thus, a 1,1 is a babysitter who has taken a first aid class whereas a 2,1 is a nonbabysitter who has taken a first aid class. This coding system is illustrated in the following screen capture of the Data Editor.

File Edit View Data Transform Analyze Graphs Utilities Add-ons Win

	BabysitterNonbabysitter	FirstAidNoFirst...	var	var
1	1	1		
2	1	1		
3	1	1		
4	1	1		
5	1	1		
6	1	1		
7	1	1		
8	1	1		
9	1	1		
10	1	1		
11	1	1		
12	1	1		
13	1	2		
14	1	2		
15	1	2		
16	1	2		
17	1	2		
18	1	2		
19	2	1		
20	2	1		
21	2	1		
22	2	1		
23	2	1		
24	2	2		
25	2	2		
26	2	2		
27	2	2		
28	2	2		
29	2	2		
30	2	2		
31	2	2		
32	2	2		
33	2	2		
34	2	2		

Please note that not all of the data are visible in the preceding screen capture, so enter the data from Table 14.2 in which there is data for 36 individuals.

Next, click on **Analyze**, **Descriptive Statistics**, and then **Crosstabs**, as indicated in the following screen capture.

This will produce the following dialog box.

Next, using the arrow keys, move one of the variables to the **R<u>o</u>w(s)** box and move the second variable to the **<u>C</u>olumn(s)** box (it does not matter which variable is entered into rows versus columns). The dialog box should now look as follows.

Click on **Statistics**, check the **Chi-square** box, and click **Continue**.

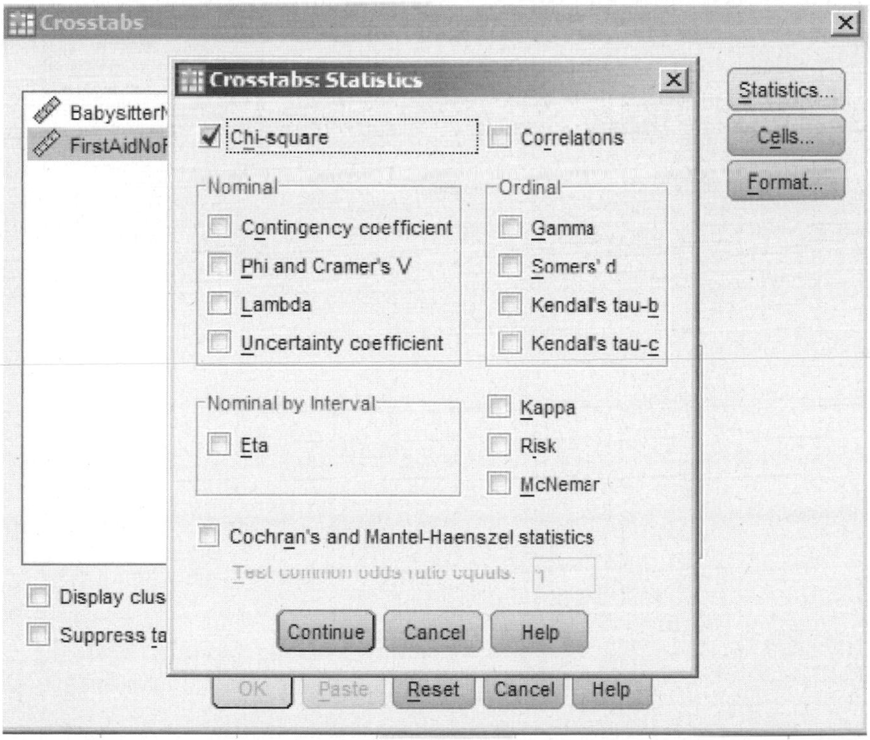

Next, click on **Cells** and check the **Observed** and **Expected** boxes as seen in the following image.

Click **Continue** and then **OK**. The output should appear as follows.

Crosstabs

Case Processing Summary

	Cases					
	Valid		Missing		Total	
	N	Percent	N	Percent	N	Percent
BabysitterNonbabysitter * FirstAidNoFirstAid	36	33.0%	73	67.0%	109	100.0%

BabysitterNonbabysitter * FirstAidNoFirstAid Crosstabulation

			FirstAidNoFirstAid		Total
			1	2	
BabysitterNonbabysitter	1	Count	12	6	18
		Expected Count	8.5	9.5	18.0
	2	Count	5	13	18
		Expected Count	8.5	9.5	18.0
Total		Count	17	19	36
		Expected Count	17.0	19.0	36.0

Chi-Square Tests

	Value	df	Asymp. Sig. (2-sided)	Exact Sig. (2-sided)	Exact Sig. (1-sided)
Pearson Chi-Square	5.461[a]	1	.019		
Continuity Correction[b]	4.012	1	.045		
Likelihood Ratio	5.611	1	.018		
Fisher's Exact Test				.044	.022
Linear-by-Linear Association	5.310	1	.021		
N of Valid Cases	36				

a. 0 cells (.0%) have expected count less than 5. The minimum expected count is 8.50.

b. Computed only for a 2x2 table

We are most interested in the data reported in the **Chi-Square Tests** box above. The chi-square test of independence is reported as **Pearson Chi-Square** by SPSS. Based on the calculations, we reject the null hypothesis. In other words, there is a significant difference between babysitters and nonbabysitters in terms of their having taken a first aid class—significantly more babysitters have taken a first aid class. If you were to report this result in APA style, it would appear as χ^2 (1, $N = 190$) = 5.461, $p = .019$.

Using the TI-84

Let's use the data from Table 14.2 to conduct the calculation using the TI-84 calculator:

1. With the calculator on, press the 2nd key followed by the **MATRIX** [X^{-1}] key.
2. Highlight **EDIT** and **1:[A]** and press **ENTER**.
3. Enter the dimensions for the matrix. Our matrix is 2×2. Press **ENTER**.
4. Enter each observed frequency from Table 14.2 followed by **ENTER**.
5. Press the **STAT** key and highlight **TESTS**.
6. Scroll down to **C: χ^2-Test** and press **ENTER**.
7. The calculator should show **Observed: [A]** and **Expected: [B]**.
8. Scroll down and highlight **CALCULATE** and press **ENTER**.

The χ^2 value of a (5.461) should appear on the screen, along with the $df = 1$ and $p = .019$.

MODULE EXERCISES

(Answers appear in Appendix.)

1. A researcher believes that the percentage of people who exercise in California is greater than the national exercise rate. The national rate is 20%. The researcher gathers a random sample of 120 individuals who live in California and finds that the number who exercise regularly is 31 out of 120. Use SPSS and the TI-84 to calculate the chi-square goodness-of-fit test.

2. A teacher believes that the percentage of students at her high school who go on to college is greater than the rate in the general population of high school students. The rate in the general population is 30%. In the most recent graduating class at her high school, the teacher found that 90 students graduated and that 40 of those went on to college. Use SPSS and the TI-84 to calculate the chi-square goodness-of-fit test.

3. You notice in your introductory psychology class that more women tend to sit up front and more men in the back. In order to determine whether this difference is significant, you collect data on the seating preferences for the students in your class. The data appear below.

	Males	Females
Front of the Room	15	27
Back of the Room	32	19

Use SPSS and the TI-84 to calculate the chi-square test of independence.

Module 15

✳

Wilcoxon Tests

In this module, we discuss two Wilcoxon tests. The *Wilcoxon rank-sum test* is similar to the independent-groups *t* test, and the *Wilcoxon matched-pairs signed-ranks test* is similar to the correlated-groups *t* test. The Wilcoxon tests, however, are nonparametric tests. As such, they use ordinal data rather than interval-ratio data and allow us to compare the medians of two populations instead of the means.

WILCOXON RANK-SUM TEST

Imagine that a fifth-grade teacher wants to compare the number of books read per term by female versus male students in her class. The distribution representing the number of books read is skewed (not normal), and thus, a nonparametric test is used. Additionally, the Wilcoxon test is based on rankings (ordinal data), and therefore, the number of books read by each student will be converted to a ranked score in order to calculate the test statistic. The teacher predicts that the girls will read more books than the boys. Thus, H_0 is that the median number of books read does not differ between girls and boys ($Md_{girls} = Md_{boys}$, or $Md_{girls} \leq Md_{boys}$), and H_a is that the median number of books read is greater for girls than for boys ($Md_{girls} > Md_{boys}$). The number of books read by each group is presented in Table 15.1.

TABLE 15.1   **Number of Books Read by Female and Male Students**
© Cengage Learning 2013

Girls	Boys
20	10
24	17
29	23
33	19
57	22
35	21

Using SPSS

To begin using SPSS to calculate this Wilcoxon rank–sum test, we must enter the data into the Data Editor. This operation is illustrated in the following screen capture of the Data Editor in which you can see that the first column is labeled Gender with females as 1s and males as 2s. The number of books read by each student is indicated in the second column.

	Gender	BooksRead	var	var	var
1	1.00	20.00			
2	1.00	24.00			
3	1.00	29.00			
4	1.00	33.00			
5	1.00	57.00			
6	1.00	35.00			
7	2.00	10.00			
8	2.00	17.00			
9	2.00	23.00			
10	2.00	19.00			
11	2.00	22.00			
12	2.00	21.00			
13					

To conduct the statistical test begin by clicking on **Analyze** then **Nonpara-metric Tests**, **Legacy Dialogs**, and finally **2 Independent Samples** as is indicated in the following screen capture.

This action will produce the following dialog box.

Gender is the **Grouping Variable**, so use the arrow key to enter it into the appropriate box. The **Test Variable List** is Books Read, so enter it into that box. With the **Grouping Variable** box highlighted, click on the **Define Groups** box to receive the dialog box presented next.

Let SPSS know how the groups are defined, in other words that you have 1s and 2s defining your groups, and then click **Continue**. Your completed dialog box should appear as follows.

Please note that the **Test Type** selected should be **Mann–Whitney U**; however, the output you receive will have both the Mann–Whitney test and the Wilcoxon test statistics. Now, click on **OK** to receive the following analysis output.

NPar Tests

Mann-Whitney Test

Ranks

	Gender	N	Mean Rank	Sum of Ranks
BooksRead	1.00	6	9.00	54.00
	2.00	6	4.00	24.00
	Total	12		

Test Statistics[b]

	BooksRead
Mann-Whitney U	3.000
Wilcoxon W	24.000
Z	-2.402
Asymp. Sig. (2-tailed)	.016
Exact Sig. [2*(1-tailed Sig.)]	.015[a]

a. Not corrected for ties.

b. Grouping Variable: Gender

Notice that in the **Test Statistics** box, both the Mann–Whitney U test and the Wilcoxon W test statistics are provided. We can see that the mean rank for girls is 9, whereas the mean rank for boys is 4. The easiest way to interpret both the Mann–Whitney and Wilcoxon tests is based on the z score provided. Based on this score, we can conclude that girls read significantly more books than boys, with $z = -2.402$, $p = .0075$ (the two–tailed significance level is provided, but because this was a one–tailed test, we divide that in half).

WILCOXON MATCHED-PAIRS
SIGNED-RANKS TEST

Imagine that the same teacher in the previous problem wants to compare the number of books read by all students (female and male) over two terms. During the first term, the teacher keeps track of how many books each student reads.

During the second term, the teacher institutes a reading reinforcement program through which students can earn prizes based on the number of books read. The number of books read by students is once again measured. As before, the distribution representing the number of books read is skewed (not normal). Thus, a nonparametric statistic is necessary. However, in this case, the design is within-participants, as two measures are taken on each student, one before the reading reinforcement program is instituted and one after the program is instituted.

Table 15.2 shows the number of books read by the students across the two terms. Notice that the number of books read during the first term represents the data used in the previous Wilcoxon rank-sum test. The teacher uses a one-tailed test and predicts that students will read more books after the reinforcement program is instituted. Thus, H_0 is that the median number of books read does not differ between the two terms ($Md_{before} = Md_{after}$, or $Md_{before} \geq Md_{after}$), and H_a is that the median number of books read is greater after the reinforcement program is instituted ($Md_{before} < Md_{after}$).

T A B L E 15.2   Number of Books Read in Each Term
© Cengage Learning 2013

Term I (No Reinforcement)	Term II (Reinforcement)
10	15
17	23
19	20
20	20
21	28
22	26
23	24
24	29
29	37
33	40
57	50
35	55

Using SPSS

To begin using SPSS to calculate this test statistic, we must enter the data into the Data Editor as illustrated in the following screen capture.

File	Edit	View	Data	Transform	Analyze	Graphs	Utilities	Add-

| 1 : Terml | 20.00 |

	Terml	Termll	var	var
1	20.00	20.00		
2	24.00	29.00		
3	29.00	37.00		
4	33.00	40.00		
5	57.00	50.00		
6	35.00	55.00		
7	10.00	15.00		
8	17.00	23.00		
9	23.00	24.00		
10	19.00	20.00		
11	22.00	26.00		
12	21.00	28.00		
13				
14				

Notice that the variables are simply labeled **TermI** and **TermII**. Next, click on **Analyze**, **Nonparametric**, **Legacy Dialogs**, and finally **2 Related Samples**, as is indicated in the following screen capture.

File	Edit	View	Data	Transform	Analyze	Graphs	Utilities	Add-ons	Window	Help

Reports ▶
Descriptive Statistics ▶
Compare Means ▶
General Linear Model ▶
Correlate ▶
Regression ▶
Classify ▶
Dimension Reduction ▶
Scale ▶
Nonparametric Tests ▶   One Sample...
Forecasting ▶   Independent Samples...
Multiple Response ▶   Related Samples...
Quality Control ▶   Legacy Dialogs ▶
ROC Curve...

Legacy Dialogs submenu:
- Chi-square...
- Binomial...
- Runs...
- 1-Sample K-S...
- 2 Independent Samples...
- K Independent Samples...
- 2 Related Samples...
- K Related Samples...

1 : Terml   20.00

	Terml	Term...	var	var	var	var	var
1	20.00						
2	24.00						
3	29.00						
4	33.00						
5	57.00						
6	35.00						
7	10.00						
8	17.00						
9	23.00						
10	19.00						
11	22.00	26.00					
12	21.00	28.00					
13							
14							
15							
16							
17							
18							

This operation will produce the following dialog box.

Two-Related-Samples Tests

Terml
Termll

Test Pairs:

Pair	Variable1	Variable2
1		

Options...

Test Type
- ☑ Wilcoxon
- ☐ Sign
- ☐ McNemar
- ☐ Marginal Homogeneity

OK   Paste   Reset   Cancel   Help

Highlight TermI and use the arrow key to move it to the **Variable1** box. Do the same for the TermII variable. In addition, make sure that **Wilcoxon** is selected in the **Test Type** box. The dialog box should appear as follows.

Next, click on the **Options** box to receive the following dialog box.

Click **Descriptive** and then **Continue**. Finally click **OK** to receive the following analysis output.

NPar Tests

Descriptive Statistics

	N	Mean	Std. Deviation	Minimum	Maximum
TermI	12	25.8333	11.96839	10.00	57.00
TermII	12	30.5833	12.43498	15.00	55.00

Wilcoxon Signed Ranks Test

Ranks

		N	Mean Rank	Sum of Ranks
TermII - TermI	Negative Ranks	1[a]	8.00	8.00
	Positive Ranks	10[b]	5.80	58.00
	Ties	1[c]		
	Total	12		

a. TermII < TermI

b. TermII > TermI

c. TermII = TermI

Test Statistics[b]

	TermII - TermI
Z	-2.229[a]
Asymp. Sig. (2-tailed)	.026

a. Based on negative ranks.

b. Wilcoxon Signed Ranks Test

We can see that the mean number of books read in TermI is 25.83, whereas the mean number of books read in TermII is 30.58. In order to conduct the Wilcoxon test, the number of books read in TermI is subtracted from the number of books read in TermII for each student. These difference scores are then ranked (an ordinal variable). We can see that there was one negative rank, 10 positive ranks, and 1 tie. Based on these results we can conclude that students read significantly more books in TermII in comparison to TermI, with $z = -2.229$, $p = .013$ (the two-tailed significance level is provided, but because this was a one-tailed test, we divide that in half).

MODULE EXERCISES

(Answers appear in Appendix.)

1. A researcher is interested in comparing the maturity level of students who
    volunteer for community service versus those who do not. The researcher
    assumes that those who complete community service will have higher
    maturity scores. Maturity scores tend to be skewed (not normally distributed).
    The maturity scores appear below. Higher scores indicate higher maturity
    levels.

No Community Service	Community Service
33	41
41	48
54	61
13	72
22	83
26	55

    Use SPSS to conduct the Wilcoxon rank-sum test.

2. Researchers at a food company are interested in how a new spaghetti
    sauce made from green tomatoes (and green in color) will compare to
    their traditional red spaghetti sauce. They are worried that the green color
    will adversely affect the tastiness scores. They randomly assign participants
    to either the green or red sauce condition. Participants indicate the tastiness
    of the sauce on a 10-point scale. Tastiness scores tend to be skewed. The
    scores appear below.

Red Sauce	Green Sauce
7	4
6	5
9	6
10	8
6	7
7	6
8	9

    Use SPSS to conduct the Wilcoxon rank-sum test.

3. Imagine that the researchers in Exercise 2 want to conduct the same study as a within-participants design. Participants rate both the green and red sauces by indicating the tastiness of the sauce on a 10-point scale. As in the previous problem, researchers are concerned that the color of the green sauce will adversely affect tastiness scores. Tastiness scores tend to be skewed. The scores appear below.

Red Sauce	Green Sauce
7	4
6	3
9	6
10	8
6	7
7	5
8	9

© Cengage Learning 2013

Use SPSS to conduct the Wilcoxon matched–pairs signed–ranks test.

Appendix

✳

Answers to Module Exercises

MODULE 1

Module 1 contains no exercises.

MODULE 2

1.

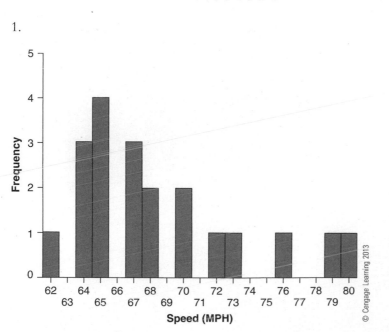

2.

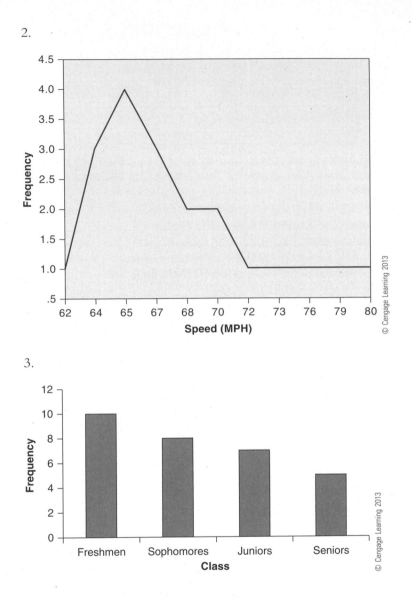

3.

MODULE 3

1. Mean = 7.3
2. Mean = 4.83
3. Mean = 6.17
4. Mean = 5.5
5. Standard Deviation = 2.74
6. Standard Deviation = 2.74

7. Standard Deviation = 27.39
8. Standard Deviation = .274
9. Standard Deviation = 273.86

MODULE 4

1. $z = +1.33$
2. $z = -1.67$
3. $z = +.50$

MODULE 5

1. a. This is a one-tailed test.
    b. $z = 2.37$
    c. Reject H_0. High school students at private high schools score significantly higher on the SAT.
2. a. This is a one-tailed test.
    b. $z = -1.59$
    c. Fail to reject H_0. The new toothpaste does not lead to significantly fewer cavities.

MODULE 6

1. a. This is a one-tailed test.
    b. $t = -3.35$
    c. Reject H_0. Those who listen to music via headphones score significantly lower on a hearing test.
2. a. This is a two-tailed test.
    b. $t = .98$
    c. Fail to reject H_0. There was not a significant difference in spatial ability between those who listen to classical music and those in the general population who do not listen to classical music.

MODULE 7

1.

t-Test: Two-Sample Assuming Equal Variances		
	Variable 1	Variable 2
Mean	19	21.28571
Variance	37.33333333	22.2381
Observations	7	7
Pooled Variance	29.78571429	
Hypothesized Mean Difference	0	
df	12	
t Stat	-0.783523371	
P(T<=t) one-tail	0.224254341	
t Critical one-tail	1.782287556	
P(T<=t) two-tail	0.448508682	
t Critical two-tail	2.17881283	

2.

t-Test: Two-Sample Assuming Equal Variances		
	Variable 1	Variable 2
Mean	6	7.555556
Variance	1	2.277778
Observations	9	9
Pooled Variance	1.638888889	
Hypothesized Mean Difference	0	
df	16	
t Stat	-2.57760893	
P(T<=t) one-tail	0.010119899	
t Critical one-tail	1.745883676	
P(T<=t) two-tail	0.020239798	
t Critical two-tail	2.119905299	

MODULE 8

1. $t(5) = -6.71$, $p = .000557$. Reject H_0. Participating in sports leads to significantly higher self-esteem scores.

2. $t(5) = 2.71$, $p = .01767$. Reject H_0. Participants had significantly higher scores on the test when they studied without music.

MODULE 9

1.

Anova: Single Factor					
SUMMARY					
Groups	Count	Sum	Average	Variance	
Column 1	5	20	4	2.5	
Column 2	5	24	4.8	6.7	
Column 3	5	53	10.6	2.3	
Column 4	5	52	10.4	2.3	
ANOVA					
Source of Variation	SS	df	MS	F	P-value	F crit
Between Groups	187.75	3	62.58333	18.1401	2.12E-05	3.238872
Within Groups	55.2	16	3.45			
Total	242.95	19				

2.

Anova: Single Factor					
SUMMARY					
Groups	Count	Sum	Average	Variance	
Column 1	5	11	2.2	0.7	
Column 2	5	14	2.8	0.7	
Column 3	5	23	4.6	1.3	
ANOVA					
Source of Variation	SS	df	MS	F	P-value	F crit
Between Groups	15.6	2	7.8	8.666667	0.004687	3.885294
Within Groups	10.8	12	0.9			
Total	26.4	14				

MODULE 10

1.

Anova: Two-Factor Without Replication						
SUMMARY	*Count*	*Sum*	*Average*	*Variance*		
Row 1	3	13	4.333333	2.333333		
Row 2	3	14	4.666667	0.333333		
Row 3	3	9	3	1		
Row 4	3	9	3	4		
Row 5	3	16	5.333333	4.333333		
Row 6	3	13	4.333333	2.333333		
Row 7	3	9	3	1		
Column 1	7	18	2.571429	0.952381		
Column 2	7	28	4	1.333333		
Column 3	7	37	5.285714	1.238095		
ANOVA						
Source of Variation	*SS*	*df*	*MS*	*F*	*P-value*	*F crit*
Rows	16.28571	6	2.714286	6.705882	0.002658	2.99612
Columns	25.80952	2	12.90476	31.88235	1.58E-05	3.885294
Error	4.857143	12	0.404762			
Total	46.95238	20				

2.

Anova: Two-Factor Without Replication				
SUMMARY	Count	Sum	Average	Variance
Row 1	3	10	3.333333	0.333333
Row 2	3	14	4.666667	0.333333
Row 3	3	9	3	1
Row 4	3	6	2	1
Row 5	3	10	3.333333	0.333333
Row 6	3	11	3.666667	0.333333
Row 7	3	7	2.333333	0.333333
Row 8	3	8	2.666667	2.333333
Row 9	3	11	3.666667	1.333333
Row 10	3	7	2.333333	0.333333
Column 1	10	24	2.4	0.933333
Column 2	10	32	3.2	0.844444
Column 3	10	37	3.7	0.9

ANOVA						
Source of Variation	SS	df	MS	F	P-value	F crit
Rows	17.36667	9	1.92963	5.158416	0.001533	2.456281
Columns	8.6	2	4.3	11.49505	0.000607	3.554557
Error	6.733333	18	0.374074			
Total	32.7	29				

MODULE 11

1.

Anova: Two-Factor With Replication						
SUMMARY	Females	Males	Total			
Low Fat						
Count	6	6	12			
Sum	19	46	65			
Average	3.166667	7.666667	5.416667			
Variance	1.366667	1.466667	6.810606			
Regular						
Count	6	6	12			
Sum	53	24	77			
Average	8.833333	4	6.416667			
Variance	1.366667	2.8	8.265152			
Total						
Count	12	12				
Sum	72	70				
Average	6	5.833333				
Variance	10	5.606061				
ANOVA						
Source of Variation	*SS*	*df*	*MS*	*F*	*P-value*	*F crit*
Sample	6	1	6	3.428571	0.0789	4.351244
Columns	0.166667	1	0.166667	0.095238	0.760812	4.351244
Interaction	130.6667	1	130.6667	74.66667	3.47E-08	4.351244
Within	35	20	1.75			
Total	171.8333	23				

2.

Anova: Two-Factor With Replication

SUMMARY	Females	Males	Total			
2 Hours						
Count	5	5	10			
Sum	59	59	118			
Average	11.8	11.8	11.8			
Variance	0.7	0.7	0.622222			
4 Hours						
Count	5	5	10			
Sum	45	45	90			
Average	9	9	9			
Variance	2	1	1.333333			
6 Hour						
Count	5	5	10			
Sum	32	33	65			
Average	6.4	6.6	6.5			
Variance	0.8	1.8	1.166667			
Total						
Count	15	15				
Sum	136	137				
Average	9.066667	9.133333				
Variance	6.209524	5.838095				

ANOVA

Source of Variation	SS	df	MS	F	P-value	F crit
Sample	140.6	2	70.3	60.25714	4.4E-10	3.402826
Columns	0.033333	1	0.033333	0.028571	0.867189	4.259677
Interaction	0.066667	2	0.033333	0.028571	0.971866	3.402826
Within	28	24	1.166667			
Total	168.7	29				

MODULE 12

1. $r = +.852$
2. IQ with psychology exam: $r = +.55$
   IQ exam with statistics exam: $r = +.68$
   Psychology exam with statistics exam: $r = +.63$

MODULE 13

1. $Y' = 1.22(X) + .41$
2. IQ and psychology exam: $Y' = .27(X) + 9.00$
    IQ and statistics exam: $Y' = .32 (X) + 4.97$
    Psychology exam and statistics exam: $Y' = .59(X) + 17.43$

MODULE 14

1.

Chi-Square Test

Frequencies

NumberExercising

	Observed N	Expected N	Residual
1.00	31	24.0	7.0
2.00	89	96.0	-7.0
Total	120		

Test Statistics

	Number Exercising
Chi-Square	2.552[a]
df	1
Asymp. Sig.	.110

a. 0 cells (.0%) have expected frequencies less than 5. The minimum expected cell frequency is 24.0.

2.

Chi-Square Test

Frequencies

GoToCollege

	Observed N	Expected N	Residual
1.00	40	27.0	13.0
2.00	50	63.0	-13.0
Total	90		

Test Statistics

	GoToCollege
Chi-Square	8.942[a]
df	1
Asymp. Sig.	.003

a. 0 cells (.0%) have expected frequencies less than 5. The minimum expected cell frequency is 27.0.

3.

Crosstabs

Case Processing Summary

	Cases					
	Valid		Missing		Total	
	N	Percent	N	Percent	N	Percent
Gender * FrontBack	93	100.0%	0	.0%	93	100.0%

Gender * FrontBack Crosstabulation

			FrontBack		Total
			1.00	2.00	
Gender	1.00	Count	15	32	47
		Expected Count	21.2	25.8	47.0
	2.00	Count	27	19	46
		Expected Count	20.8	25.2	46.0
Total		Count	42	51	93
		Expected Count	42.0	51.0	93.0

Chi-Square Tests

	Value	df	Asymp. Sig. (2-sided)	Exact Sig. (2-sided)	Exact Sig. (1-sided)
Pearson Chi-Square	6.732[a]	1	.009		
Continuity Correction[b]	5.694	1	.017		
Likelihood Ratio	6.817	1	.009		
Fisher's Exact Test				.013	.008
Linear-by-Linear Association	6.660	1	.010		
N of Valid Cases	93				

a. 0 cells (.0%) have expected count less than 5. The minimum expected count is 20.77.
b. Computed only for a 2x2 table

MODULE 15

1.

Mann-Whitney Test

Ranks

	CommunityService	N	Mean Rank	Sum of Ranks
MaturityScores	1.00	6	3.92	23.50
	2.00	6	9.08	54.50
	Total	12		

Test Statistics[b]

	MaturityScores
Mann-Whitney U	2.500
Wilcoxon W	23.500
Z	-2.486
Asymp. Sig. (2-tailed)	.013
Exact Sig. [2*(1-tailed Sig.)]	.009[a]

a. Not corrected for ties.

b. Grouping Variable: CommunityService

2.

Mann-Whitney Test

Ranks

	SauceType	N	Mean Rank	Sum of Ranks
TastinessScore	1.00	7	8.86	62.00
	2.00	7	6.14	43.00
	Total	14		

Test Statistics[b]

	Tastiness Score
Mann-Whitney U	15.000
Wilcoxon W	43.000
Z	-1.236
Asymp. Sig. (2-tailed)	.217
Exact Sig. [2*(1-tailed Sig.)]	.259[a]

a. Not corrected for ties.

b. Grouping Variable: SauceType

3.

Wilcoxon Signed Ranks Test

Ranks

		N	Mean Rank	Sum of Ranks
GreenSauce - RedSauce	Negative Ranks	5[a]	4.60	23.00
	Positive Ranks	2[b]	2.50	5.00
	Ties	0[c]		
	Total	7		

a. GreenSauce < RedSauce

b. GreenSauce > RedSauce

c. GreenSauce = RedSauce

Test Statistics[b]

	GreenSauce - RedSauce
Z	-1.552[a]
Asymp. Sig. (2-tailed)	.121

a. Based on positive ranks.

b. Wilcoxon Signed Ranks Test

Index